MW00467290

CALIFORNIA NATURAL HISTORY GUIDES

SHARKS, RAYS, AND CHIMAERAS OF CALIFORNIA

California Natural History Guides

Phyllis M. Faber and Bruce M. Pavlik, General Editors

SHARKS, RAYS, and CHIMAERAS of California

David A. Ebert

Illustrated by Mathew D. Squillante

UNIVERSITY OF CALIFORNIA PRESS

Berkeley   Los Angeles   London

In memory of my sister, Margaret

California Natural History Guides No. 71

University of California Press
Berkeley and Los Angeles, California

University of California Press, Ltd.
London, England

Library of Congress Cataloging-in-Publication Data

Ebert, David A.
    Sharks, rays, and chimaeras of California / David A. Ebert; illustrations by
    Mathew D. Squillante.
        p.   cm.
    Includes bibliographical references (p.) and indexes.
    ISBN 0–520-22265-2 (hardcover : alk. paper) — ISBN 0–520-23484-7 (pbk. : alk.
paper)
    1. Chondrichthyes—California.   I. Title.

QL638.6 E24   2003
597.3'17743—dc21   2002032079

Manufactured in China
10   09   08   07   06   05   04   03
10   9   8   7   6   5   4   3   2   1

The publisher gratefully acknowledges the generous
contributions to this book provided by

the Moore Family Foundation
Richard and Rhoda Goldman Fund
and
the General Endowment Fund of the
University of California Press Associates

CONTENTS

FOREWORD

Sharks, Rays, and Chimaeras of California is the sort of book that I wish I'd been able to write when I was living in the San Francisco Bay Area. One of my former doctoral students has done so instead, producing a work that is a major step forward in the annals of shark research in California. The scientific and semipopular literature on the sharklike fishes of California is remarkably scattered, limited, and uneven, and many of the species are sketchily known taxonomically, morphologically, and biologically. This is the first work dedicated to the subject in a half century, and it far exceeds its antecedents, presenting much new biological data drawn from many new records.

In California, as elsewhere in the world, research on sharklike fishes has focused on commercially important species such as the Soupfin, Shortfin Mako, Common Thresher, and Pacific Angel Sharks. Great White Sharks have also been the subject of much research attention as a result of the media hype around "attacks" on humans; I remember cringing at every headline shouting "Shark Attack" during my childhood in San Francisco, and the legendary status of the species was only intensified by the film *Jaws,* which obscured the reality of Great White Shark behavior for two decades or more. In the meantime, lesser known species have been neglected.

Here in California, where the Great White Shark was made into a science fiction superstar by Universal Studios, there should be great interest in the study of cartilaginous fishes. But jobs for such work have been few and far between, and the number of professional shark researchers based in California remains small and has diminished in recent decades. Much work, including detailed biological studies of many species, remains to be done.

And David Ebert is the man for the job. A native Californian, Dave completed his master's degree on California cowsharks at Moss Landing Marine Laboratories under Gregor Cailliet. He is also the world's leading authority on cow and frilled sharks, having helped to revise the classification of this order on a worldwide scale during his doctoral work at Rhodes University in Grahamstown, South Africa, and at the Shark Research Center at the South African Museum in Cape Town.

Sharks, Rays, and Chimaeras of California invites comparison with Peter Last and John Stevens' *Sharks and Rays of Australia* (1994), and particularly with the *Guide to the Sharks and Rays of Southern Africa* (1989), which Dave coauthored with me and Malcolm Smale. The Australian work, dealing with a massive and poorly known cartilaginous fish fauna of over 300 species, is primarily a field and taxonomic guide with limited natural history data. The southern Africa guide is more like the present book, and except that it covers over twice as many species, it is far more limited and simple in its tight field-guide format and has far less information on the morphology and natural history of individual species.

Sharks, Rays, and Chimaeras of California, in terms of its depth and breadth of coverage, is an essential and definitive reference to California cartilaginous fishes.

L. J. V. Compagno

PREFACE

The idea for this book grew out of years of working with anglers, commercial fishermen, colleagues, and amateur naturalists who had trouble identifying many of the various local sharks and rays. I often found myself trying to obtain accurate descriptions from people with little to no knowledge of what characteristics to look for on a shark. The rays, particularly the skates, were even more problematic. Although people have some idea of what skates look like, even many professional ichthyologists have difficulty identifying them.

During my doctoral research in South Africa, I often expressed frustration to my friend and supervisor, Leonard Compagno, about the fact that we were losing potentially valuable specimens because of the fishing community's lack of knowledge in identifying the various cartilaginous fishes they encountered. Many of these species were either poorly known or entirely new to the scientific community. Although the local fishing communities had their own ways of differentiating the various species from each other, it was often difficult to discern from their descriptions whether a particular specimen would be of interest to us. Although putting together a makeshift identification guide helped somewhat, it was apparent that much could still be done to assist those who spent a considerable amount of time around the sea in identifying the various local cartilaginous fishes. We eventually developed a field guide with color illustrations for each of the sharks, rays, and chimaeras known at that time to occur in the waters of southern Africa (Compagno et al. 1989). The book was very well received by anglers, commercial fishermen, divers, marine biologists, and naturalists, among others, who used it to identify and in many cases collect specimens previously unknown. Returning to my native California, I continued my research on chondrichthyans and again found that because of a lack of knowledge, it was difficult to get people to save rare or unusual specimens since most didn't know what was rare or unusual. To facilitate a means to identify these animals, I embarked on a similar project to develop an easy-to-use field guide to our local fauna. The results of this effort are contained in this book.

To prepare this book, I conducted extensive archival literature searches, examined many museum specimens, called on the generosity of colleagues to provide unpublished data from their own research, and reviewed extensive field notes that I have kept over the years. Much has changed in the 50 plus years since the last publication devoted exclusively to California's chondrichthyan fauna: New classifications have been erected, new species have been added, and scientific names have changed. Some of the groups, such as the catsharks and skates, are poorly known and most likely will require significant future revisions which may yield additional new species than those currently known in California.

My own interest in cartilaginous fishes began when I was about five years old and my parents gave me a book on sharks. Since then, and with continued encouragement from my parents, I have turned a youthful fascination into a lifelong passion. I hope that through this book I can pass along my enthusiasm to other young people and to those of us who are young at heart.

David Ebert
September 2002

ACKNOWLEDGMENTS

Author's Acknowledgments

This book would not have been possible without the generous assistance of friends and colleagues. I thank Rachel Alexander, Tim Athens, Felipe Barreto, Joe Bizzarro, George Burgess, Dave Catania, Dominique Didier-Daget, Manny Ezcurra, Ken Goldman, Scott Greenwald, Randy Hamilton, Nancy Kohler, Chris Lowe, Henry Mollet, Lisa Natanson, John O'Sullivan, Joan Parker, Ross Pounds, Ron Russo, Jeff Seigel, Wade Smith, Tim Thomas, and Gilbert van Dykhuizen.

I especially acknowledge three individuals who have been instrumental in providing guidance throughout my career. They are Greg Cailliet of Moss Landing Marine Laboratories, my master's thesis supervisor; Leonard Compagno of the Shark Research Center, South African Museum, who supervised my Ph.D. thesis; and my father, who has been an inspiration to me throughout my life.

Words cannot adequately express my appreciation to both my parents, Earl and Peggy Ebert, for all their years of encouragement, support, and love.

Illustrator's Acknowledgments

I thank the following people for their invaluable support during this project: David Blackburn, Jason and Stephanie Blackburn, Ann Caudle, Jenny Keller, Keiko Okushi, Jason and Melissa Robinson, and Lesley Squillante.

Special thanks are extended to the University of California, Santa Cruz, Science Illustration Program, which provided the education and training so pivotal to the creation of the artwork of this book.

INTRODUCTION

Early Studies of California Sharks

This field guide to California's sharks, rays, and chimaeras—collectively known as the chondrichthyans or cartilaginous fishes—is the first in more than 50 years devoted exclusively to this fascinating group. The last comprehensive guide to California's chondrichthyan fauna was published in the California Department of Fish and Game's Fish Bulletin series by Phil Roedel and William Ripley (no. 75, 1950). Since then, several guides to the eastern North Pacific fishes have been published (Eschmeyer et al. 1983; Hart 1973; Miller and Lea 1972), but their usefulness has been hampered by their cursory treatment of the fauna, inadequate illustrations, or incomplete identification keys. The sharks have received the most attention, to the virtual exclusion of their flattened relatives, the rays or batoids, and the even more poorly known chimaeras, also called ratfishes or silver sharks. Since the publication by Roedel and Ripley, 14 sharks, five rays, and two chimaeras have been added to the California fauna bringing the current totals to 43 sharks, 22 rays, and three chimaeras.

California has a long and rich historical tradition of contributing to our knowledge of chondrichthyans. The first species described off its coast was the White-spotted Chimaera (Lay and Bennett 1839), using a specimen taken from Monterey Bay. Directed studies on California's fish fauna, including its chondrichthyans, were initiated during the Mexican Boundary and Pacific Railroad surveys of the 1850s. Charles Girard, of the Smithsonian Institute, and William Ayres, one of the first physicians to come west following the California gold rush, were instrumental in describing several new chondrichthyan species. Girard published the first faunal account as part of the Pacific Railroad surveys in 1858. Ayres, who settled in San Francisco, became California's first resident ichthyologist and one of the founders of the California Academy of Sciences.

North American ichthyology entered its golden age with the arrival of David Starr Jordan at Stanford University. Jordan dominated the field in the late nineteenth century, and many current ichthyologists, including this author, can trace their academic roots back to him. Among Jordan's more important works were his *Synopsis of the Fishes of North America* (Jordan and Gilbert 1883a) and *The Fishes of North and Middle America* (Jordan and

Evermann 1896), each of which added considerably to the knowledge of our chondrichthyan fauna.

In the early part of the twentieth century continued effort was directed toward identifying and documenting the Pacific coast fauna. Samuel Garman's classic *The Plagiostoma—Sharks, Skates, and Rays* (1913) included revised or new descriptions for many of the sharks and rays; J. Frank Daniel's *The Elasmobranch Fishes* (1934) contributed much to our understanding of chondrichthyan anatomy; and Edwin Starks (1917, 1918) produced one of the first identification guides focusing exclusively on California's known chondrichthyan fauna, which at that time encompassed 18 sharks, 13 rays, and the White-spotted Chimaera. In *The Sharks of California*, Starks encouraged people to catch and eat sharks to reduce their numbers so that commercially important fish and shellfish species might be saved, in marked contrast to our current knowledge that these fishes are highly vulnerable to overexploitation. With his *Sharks and Rays of California* (1935), Lionel Walford added six more species to the known fauna. Walford provided a relatively simple means of distinguishing the different kinds of sharks and rays in California waters, in essence writing the first true field guide on the subject.

The publication of Walford's guide marked a shift in the direction of research efforts. Major shark fisheries were beginning to develop in California as part of an emerging shark-liver oil boom, and as the need grew to develop and implement a fisheries management program, the international hub for ichthyological research shifted from Stanford University to the California Department of Fish and Game. One of the more significant works of the following decade was William Ripley's *Biology of the Soupfin Shark* (1946), on the single most commercially important shark species at that time. The culmination of the Department's research was Roedel and Ripley's *Sharks and Rays of California* (1950), which served as a field guide to all of the sharks and rays known in the fauna at that time.

At about the time the Soupfin Shark fishery collapsed in the mid-1940s, when the vitamin oils extracted from Soupfin livers were first being created synthetically, the direction of California shark research changed once again, this time toward studying the rare phenomenon of shark attack. From the 1950s through the early 1970s, considerable research focused on why sharks attack in an effort to develop possible preventive measures. Much funding for this research came from the U.S. Navy, Office of Naval Research, in response to widely publicized World War II stories about downed aviators and sailors who encountered sharks while awaiting rescue. Despite this narrow focus on attack behavior, this era saw the publication of several significant works on eastern Pacific sharks, including a review of requiem sharks of the eastern Pacific by Rosenblatt and Baldwin (1958), a more general guide to sharks of the eastern Pacific by Kato et al. (1967), and a guide to California's coastal marine fishes by Miller and Lea (1972).

In the mid-1970s, following the *Jaws* movie phenomenon, shark fishing became important again, the object this time being food for human consumption. In addition to commercial fisheries for selected species, recreational fisheries emerged as an important presence, especially in southern California and in San Francisco Bay. This trend continued through the remainder of the twentieth century.

Research into chondrichthyans during the past quarter century has not lacked for effort. What has lagged have been identification, description, general biology, and natural history studies of some of the common but lesser known species. Several species of catsharks and skates, and at least one chimaera, have been awaiting formal description for at least three decades. This is unfortunate as some of these species are taken in considerable numbers, but fisheries biologists either misidentify them or lump them into a catchall category with no further information. The impact of fisheries on the populations of these species, and the impact their loss may have on the ecosystem, is largely unknown. The skates—which are so poorly known that most people, including some professional ichthyologists, are unable to identify them—are particularly troubling.

This guide was designed for amateur naturalists, anglers, and divers as well as the professional ichthyologist, and for anyone who has an interest in these fascinating fishes. Unlike previous guides to California's cartilaginous fish fauna, most of which tend to include only those species commonly seen, all 65 of the known species as well as three others believed to occur in California waters are described and illustrated in color. Because most species lose their colorful appearance soon after death, and several of the rarer species are known only from preserved museum specimens, which tend toward a uniform brown or gray, people often don't realize how colorful many of these species are. Geographically the guide reaches from the innermost shallow bays and estuaries to 500 miles offshore.

The introductory sections on chondrichthyan biology, ecology, and diversity are intended only as a brief overview, as other contemporary authors have treated these subjects in more detail. The guide is primarily intended to be useful in the field, where users can simply match the illustration to the specimen or work through the identification keys to each species. Once a tentative identification has been made, the reader can turn to the species description and natural history notes.

California's Marine Environment

The California coastline (map 1), which extends over 1,100 miles in a north–south direction from the Mexican border (32 degrees N latitude) to the Oregon border (42 degrees N latitude), is composed of three major geographic regions: northern California (the Oregon border to San Francisco), central California (San Francisco to Point Conception), and southern California (Point Conception to the Mexican border). The general flora and fauna shift from warm-temperate water species in the south to cold-temperate water species in the north, whereas the transition of the elasmobranch fauna seems to occur in the central California region, where warm-temperate water species give way to cool-temperate nearshore species. Species such as the Pacific Sleeper Shark, some skates, and the White-spotted Ratfish, which occur in deep water in southern California, occur in relatively shallower water in the northern part of the state.

The temperature of the water along the California coast is influenced by the California Current, a major surface current that flows north to south, from the Gulf of Alaska, down the coastline to Point Conception, where it veers offshore. The area south of Point Conception where the coastline bends eastward is known as the Southern California Bight and is less affected by the California Current than the waters to the north. The less dominant Davidson Current brings warmer water from the south along the coastline to the north influencing the Southern California Bight and the Channel Islands. A seasonal fluctuation in the movement

Map 1. The California coast.

of these currents combined with strong seasonal winds causes an upwelling of cold, nutrient-rich water along the coast, an effect most prominent in central and northern California during the spring and summer.

Water temperature is an important factor influencing the distribution of California's cartilaginous fish fauna. The gradient of surface water temperatures ranges from approximately 68 degrees F in the south during summer to about 48 degrees F in the north during winter and spring, with Point Conception being the major area of demarcation. South of Point Conception the water temperature is usually 4 to 11 degrees F warmer than to the north. Temperature changes created by surface winds also occur throughout the water column, forming stratified layers of warm and cold water known as a thermocline.

The seasonal movement of warm or cool water masses influences the movements of many species of cartilaginous fish. In summer warm water from the Davidson Current pushes northward increasing the water temperature, bringing warmer water species into California waters. Conversely, in winter and spring cooler water masses lower the water temperature, bringing cold-temperate species into our waters. Occasionally rapid changes occur in these warm- or cold-water masses, causing a localized mortality in temperature-sensitive populations. Young dead or dying cold-water salmon sharks occasionally wash ashore along the beaches of central and southern California following a rapid increase in water temperature. Warm-water species may also be affected by a rapid cooling of the water.

California is subject to an unusual phenomenon that at times can dramatically alter weather patterns with an associated increase in water temperature. El Niño (Spanish meaning The Child, in reference to the Christ Child), which occurs every two to seven years for unknown reasons, can range from mild to quite severe, as occurred from 1982 to 1983, when the state experienced intense winter storms. During or immediately preceding an El Niño event at least 16 species of California's known cartilaginous fish fauna were first reported in our waters, including the Pelagic Thresher Shark and Manta Rays (seen around the Channel Islands). Both species are typically found much farther south off Baja and in the Sea of Cortez. As opposed to El Niño, La Niña brings cooler water to our area and thus a more cold-temperate water fauna.

Oceanographic conditions, especially water temperature, have exhibited major cyclic fluctuations over the past several thousand years, with changes in the average water temperature occurring every few hundred years. Historical evidence indicates that as recently as the mid-1800s the coastal waters of central and southern California reflected an environment that was more warm-temperate to subtropical than is seen today. Conversely, 400 years ago, cold-temperate species now rarely seen south of central California inhabited the coast as far south as the tip of Baja. This may explain why several seemingly cool-temperate species such as the Sevengill Shark and Leopard Shark have apparently isolated populations in the northern Gulf of California.

Classification of Cartilaginous Fishes

All living organisms are grouped into a hierarchy of categories from broad down to the most specific. The basic groups, in descending order, are kingdom, phylum, class, order, family, genus, and species (table 1). Each higher level group contains one or more members of the group below it, so a class would contain one or more orders, an order one or more families, and so on down to the species level. A single scientific name is assigned to the members of each group, except at the species level, at which members are given a binomial name (*bi* meaning two and *nomial* meaning name). The scientific name is made up of a given genus and species name, both of which are usually *italicized* or underlined.

Scientific names are often followed by the name of the person who originally described the family, genus, or species, and the year in which it was first described. If the species was subsequently placed into a different genus, the person's name is put in parentheses. The Sevengill Shark, for example, was originally described by François Peron in 1807 as *Squalus cepedianus,* but was later placed into a new genus, *Notorynchus,* by William Ayres (1855). Thus, the proper citation is *Notorynchus cepedianus* (Peron, 1807).

The person's name and date of description allow researchers to trace the taxonomic history of a specific species to determine

TABLE 1. Classification for the Sevengill Shark*

Kingdom	Animalia	All Animals
Phylum	Chordata	Animals with Backbones
Class	Chondrichthyes	Cartilaginous Fishes
Order	Hexanchiformes	Cow and Frilled Sharks
Family	Hexanchidae Gray, 1851	Cowsharks
Genus	*Notorynchus* Ayres, 1855	Sevengill Shark
Species	*Notorynchus cepedianus* (Peron, 1807)	Sevengill Shark

*The classification is used to demonstrate the basic hierarchical structure of scientific names. Common names are given in the right-hand column.

whether it was previously described or is undescribed. If two species are found to be the same, the oldest description has precedence over the newer name, which then becomes invalid (commonly referred to as a junior synonym). The species names for many of the cartilaginous fishes from California waters were originally classified under different names in earlier publications.

In each species account in this guide, the full scientific name appears at the beginning of the nomenclature subsection.

What Are Cartilaginous Fishes?

Among the five vertebrate groups, the fishes are the largest and most diverse with over 25,000 species described. By comparison, the other four vertebrate groups—amphibians, reptiles, birds, and mammals—collectively have less than 20,000 known species. The fishes can be subdivided into two distinct groups: those with a bony skeleton, known as the bony fishes or teleosts, and those with a cartilaginous skeleton, referred to as cartilaginous fishes.

All living sharks, rays, and chimaeras belong to the class Chondrichthyes (Greek, *chondro* meaning cartilage and *ichthos* meaning fish), a group of aquatic, gill-breathing, finned vertebrates. In contrast to the bony fishes or class Osteichthyes (Greek, *osteos* meaning bone and *ichthos* meaning fish), these fishes have a simplified internal cartilaginous skeleton and lack true bone. Other distinguishing characteristics of the chondrichthyans include fins without bony rays, a true upper and lower jaw, and nostrils on the underside of the head. Their teeth are typically inconspicuous transverse rows or fused tooth plates, and they are continuously being replaced from inside the mouth. Cartilaginous fishes have no bony plates on the head; their scales appear in the form of small, toothlike dermal denticles known as placoid scales; and they have internal fertilization.

The class Chondrichthyes can be subdivided into two major groups, the large subclass Elasmobranchii (*elasmo* meaning plates and *branchii* meaning gills), which includes several groups of fossil sharks and all of the modern living sharks and rays, and the small subclass Holocephali (*holo* meaning entire and *cephali* meaning head), which includes all of the chimaeras. The ordinal classification used here follows Compagno (2001) but recognizes that there is still some disagreement among modern systematists (scientists who study the classification of organisms) regarding the definition of these higher groups. There are currently ten recognized orders of "sharks," a term which in the broad sense includes the rays and chimaeras. Eight of these orders comprise the "typical" sharklike fishes while the rays and chimaeras are each in their own order (table 2). For simplicity, the term "shark" will be used when referring to "typical" sharks, the term "ray" when referring to the

TABLE 2. Approximate Numbers of Families, Genera, and Species as of 31 October 2001*

Orders	Number of Families	Number of Genera	Number of Species
Hexanchiformes	2	4	5
Squaliformes	7	23	101
Pristiophoriformes	1	2	5
Squatiniformes	1	1	15
Heterodontiformes	1	1	8
Orectolobiformes	7	14	35
Lamniformes	7	10	16
Carcharhiniformes	8	49	225
Rajiformes	22	71	543
Chimaeriformes	3	6	35

*The numbers are based on the author's own database.

raylike sharks (order Rajiformes) and the term "chimaera" when referring to the chimaeras (order Chimaeriformes).

The elasmobranchs are the dominant chondrichthyan group with approximately 56 families representing 96 percent of the species; the remaining 4 percent, which includes three families, are the chimaeras. In higher taxonomic groups (genera and above) the sharks are more diverse than the rays, but among all elasmobranch species there are more rays (57 percent) than sharks (43 percent). Worldwide approximately 988 species of cartilaginous fishes have been described, with another 150 or more awaiting formal description by researchers.

What Is a Shark?

Sharks are cartilaginous fishes with a cylindrical or flattened body, five to seven paired gill openings on each side of the head, a large caudal fin, and one or two dorsal fins, with or without erect,

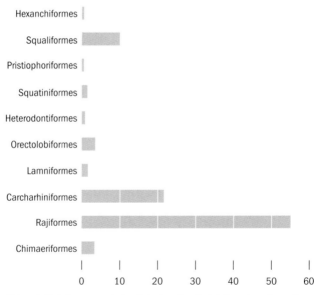

Figure 1. Relative percentage of shark orders by total number of species.

immovable spines. The moderate-sized pectoral fins are not attached to the head above the gill openings. An anal fin may be present. Lacking a swim bladder, sharks have a liver that helps them achieve neutral buoyancy.

What we think of as "typical" sharks are divided into eight major orders, each of which is easily distinguished by specific external characteristics. Of these, the largest group is the ground sharks (order Carcharhiniformes), representing about 24 percent of the shark families and about 55 percent of all shark species (fig. 1). In fact, approximately 24 percent of all elasmobranchs are members of the Carcharhiniformes. Of elasmobranchs only the rays are larger in total number of species with 543. Eight orders of shark contain 34 families and over 100 genera, and cumulatively contain at least 410 or more described species, with another 60 or so awaiting formal description. The California shark fauna is extensive, with representatives from 20 families, 30 to 31 genera, and 40 to 43 species.

What Is a Ray?

The rays, also known as batoids, are flattened or "winged" sharks whose pectoral fins expand forward and are fused to the sides of the head over the gill openings, which are on the underside of the head. Rays have a short, flat body; five or six paired gill openings; and a tail that varies from large, thick, and sharklike to slender and whiplike. One or two dorsal fins may be present but always lack a fin spine. Some species, particularly the stingrays, have tail spines, whereas most of the skates have enlarged thorns on the back and tail. Rays lack an anal fin. The pectoral fins, the main propulsive organ in rays, are greatly modified in some of the more specialized species, such as the stingrays, which have a slender whiplike tail, and the skates, which may have a small caudal fin.

The rays are the largest elasmobranch order with 22 families, 71 genera, and over 540 described species. Of ray species, the skates comprise over 44 percent (a number likely to increase as new species are described), the whiptail rays over 34 percent, the guitarfishes 11 percent, the electric rays nine percent, and the sawfishes and sharkfin guitarfishes less than one percent. The California ray fauna is represented by 10 families, 14 genera, and at least 22 species.

What Is a Chimaera?

The chimaeras are compressed, often silvery cartilaginous fishes differing from the elasmobranchs in having four pairs of gill openings, all protected by a soft gill cover with a single pair of external gill openings. Chimaeras lack the dermal denticles found in sharks and rays. Their teeth are fused into three pairs of ever growing tooth plates similar in appearance to rodent incisors, hence the common names ratfish or rabbitfish for some species. The first dorsal fin always has a spine, which can be erect or depressed. Male chimaeras have claspers on the pelvic fins, a pair of claspers (prepelvic tenacula) in front of the pelvic fins, and a single clasper (frontal tenaculum) on the forehead. Each clasper has hooklike denticles that help the male hold the female during copulation. Chimaeras propel themselves by large, fan-shaped pectoral fins.

Chimaeras have one order, three families, six genera, and over 35 described species, with as many as 15 or more species awaiting formal description. The fossil record indicates that in the past the chimaeras were far more abundant than they are today. The California chimaera fauna is represented by two families, two genera, and two described species. There is one and perhaps two undescribed species found in very deep water.

Distribution

There are two main faunal components of California's cartilaginous fishes: a cold-temperate fauna north of Point Conception and a warm-temperate fauna south of Point Conception, both heavily influenced by the prevailing surface water temperatures. The number of families, genera, and species increases as you move from northern to southern California (map 2). The southern California area appears to be a transitional zone between the warm- and cold-temperate regimes. The fauna south of Point Conception is remarkably similar to the Mexican fauna at the family level, with all but two shark and three batoid families found in both regions. Conversely the

Oregon, Washington, British Columbia, Gulf of Alaska

	Families	Genera	Species
Sharks	11	16	16
Batoids	4–5	4–5	11–12
Chimaeras	1	1	1

California

	Families	Genera	Species
Sharks	20	30–31	40–43
Batoids	10	13	22
Chimaeras	2	2	2

Mexico

	Families	Genera	Species
Sharks	21	37	61–67
Batoids	11	19	42
Chimaeras	2	3	4

Worldwide

	Families	Genera	Species
Sharks	34	104	410
Batoids	22	71	543
Chimaeras	3	6	35

Map 2. The approximate number of described chondrichthyan orders, families, genera, and species in the eastern North Pacific and worldwide. These numbers do not include those species awaiting formal description.

fauna north of Point Conception is similar to the Alaskan fauna, with most of the families also represented in the Gulf of Alaska. The most significant changes are in the species composition. The skates, in particular, become the prominent batoid group north of Point Conception with the whiptail rays becoming scarce.

The California shark fauna is composed of warm- to cool-temperate species. Most (71 percent) are fairly wide ranging, with 17 percent endemic to the eastern North Pacific; seven percent are antitropical, as they are also found in the temperate waters of Chile and Peru, and five percent are considered

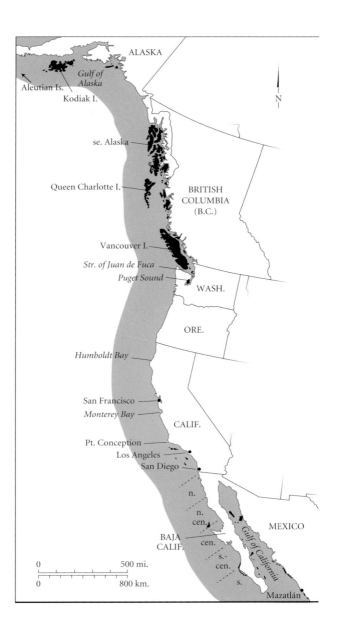

ALASKA

Gulf of
Alaska

Aleutian Is.
Kodiak I.

se. Alaska

Queen Charlotte I.

BRITISH
COLUMBIA
(B.C.)

Vancouver I.

Str. of Juan de Fuca

Puget Sound

WASH.

ORE.

Humboldt Bay

San Francisco

Monterey Bay

CALIF.

Pt. Conception

Los Angeles

San Diego

n.

n.
cen.

cen.

s.-
cen.

s.

BAJA
CALIF.

MEXICO

Gulf of California

Mazatlán

0   500 mi.

0   800 km.

N

tropical. There are no shark species found only in California waters. By comparison, of the shark fauna 48 percent of Australian species and 30 percent of southern African species are endemic.

California's batoid fauna is composed of two distinct groups: a cold-temperate skate fauna that occurs in increasingly deeper water south of Point Conception and a warmer temperate component composed of nearshore species of the families Dasyatidae, Myliobatidae, Rhinobatidae, and Urolophidae, among others. In contrast to the sharks, 60 percent of the batoids are endemic to the eastern North Pacific. Only 20 percent of the species are considered wide ranging, with five percent being antitropical and 15 percent having a tropical influence. Of the rays, 73 percent of Australian fauna and 38 percent of southern African fauna are endemic.

The California chimaera fauna is composed of two families and two genera, with two species and a third as yet undescribed species found only in very deep water. Of the two described species, one is endemic to the eastern North Pacific and the other is fairly wide ranging. Slightly more than 50 percent of the chimaeras found in Australia and about 33 percent of those found in southern Africa are endemic.

General Biology

Reproduction

Elasmobranchs undergo an elaborate and complex courtship behavior, beginning when the male repeatedly bites the female, usually around the pectoral, pelvic, and anal fins, along the abdomen, and around the gill openings. This continues until the female is receptive. Unreceptive, females may bite or snap at the male to discourage him. The skin of most females has evolved to be slightly thicker than that of the males as protection from this behavior. Larger males usually grasp the female around the gill openings or on the pectoral fin and lie side by side from head to vent with the male's body curving around the female's body so that the male can insert the claspers. One or two claspers may be inserted depending on the species. In smaller sharks, particularly catsharks, the male wraps himself around the female without biting on the gill openings or pectoral fins. During copulation the pair either sinks slowly to the bottom, gently swims together, or lies on the bottom. What induces courtship behavior is unknown as there are very few instances of actual observations. However, as in other animal groups, pheromones may play an important role in determining whether a female will be receptive.

Cartilaginous fishes have a complex and involved life cycle (fig. 2). Fertilization is internal and the mode of reproduction varies from oviparous or egg laying to viviparous or live bearing, whereby the young are nurtured internally by an independent yolk supply (ovoviviparous or aplacental viviparity) or by the mother (viviparous or placental viviparity). Male cartilaginous fishes have paired claspers located on the pelvic fins, which are the intromittent organ for fertilization. During mating, one or both claspers may be inserted into the female's cloaca, secured with spurs, hooks, or sharpened terminal edges designed to splay open after insertion.

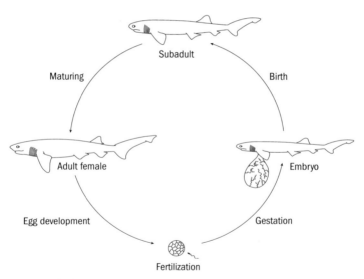

Figure 2. Generalized chondrichthyan life cycle as depicted by the Sevengill Shark.

Oviparous or egg-laying chondrichthyans usually deposit their egg cases on the bottom in mud, sand, or on rocky and coral reefs. The egg cases are usually purselike, have a conical or spindle-like shape, and have horns, tendrils, or spiral flanges to help wedge or otherwise anchor them to the bottom (fig. 3). Some species have nesting sites that are repeatedly used by several females. At least one group, the Horn Sharks, is known to pick up their egg cases with their mouths after laying them and carefully place them in an appropriate nesting site. Little information exists on incubation time for egg cases, although species reared in captivity seem to require a longer incubation period than those reared in the wild. Once born, the young immediately begin to feed on their own. There does not appear to be any maternal or paternal care for the eggs once they are laid and in fact some shark males feed on the egg cases of their own species. Besides sharks, some species of marine snails (gastropods) attack the egg cases by drilling holes in them and sucking out the yolk. Approximately 40 percent of sharks, 44 percent of rays, and all of the chimaeras lay eggs.

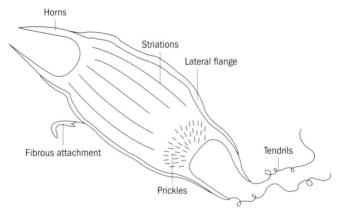

Figure 3. General illustration of chondrichthyan egg case terminology.

Viviparous (live-bearing) elasmobranchs nourish the developing embryo in one of two ways: either with or without placental attachment (fig. 4). In aplacental viviparity the developing embryo, not attached to the uterine wall, relies primarily on a large yolk sac for nourishment. Once the yolk sac has been exhausted the embryo is near full term and birth occurs shortly thereafter. In some of the more advanced rays, particularly the stingrays, the nourishment supply to the embryo is enhanced by secretion of a fluid milky nutrient through villilike extensions (trophonemata) of the uterine wall. This may in part explain why rays in general have a shorter gestation period and small litters with relatively large newborns. A rather bizarre twist to this reproductive mode, found mainly in mackerel sharks and false catsharks (family Pseudotriakidae), is that of oophagy (egg eating) and intrauterine cannibalism (embryo eating). Oophagous development is similar to that of other aplacental elasmobranchs except that when their yolk sac is used up the developing embryos begin to actively feed on ovulating eggs within the uterus. A more voracious form of intrauterine cannibalism (referred to as adelphophagy) occurs when the largest developing embryo actively attacks, kills, and eats all of the other smaller embryos within the uterus before feeding on the ovulating eggs. In sharks that exhibit intrauterine cannibalism no more than a single embryo per uterus usually survives unless its siblings are about the same size.

In viviparous or placental viviparous sharks (rays do not exhibit this mode of reproduction) the yolk supply is consumed early in the development of the embryo, at which time it becomes connected to the uterine wall of the mother, forming a yolk sac placenta. This is analogous to the mammalian placenta and serves to transfer nutrients from the mother directly to the embryos. Only 10 percent of all living sharks exhibit placental viviparity.

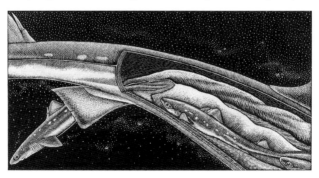

Figure 4. Viviparous elasmobranchs such as this Spiny Dogfish bear their young live, which enter the world as miniature versions of the adults.

Live-bearing elasmobranchs typically have a long gestation period, ranging from several months to over three years for at least one species. The litter size can range from one or two to at least 300 or more depending on the species. Most live-bearing elasmobranchs produce on average between two and 20 young per litter. Because the length at birth of some species may be 1 m or more, newborn sharks have very few predators other than larger sharks.

Migratory Patterns

Many cartilaginous fishes exhibit fairly complex migratory patterns associated with their life history cycle. For example, before "pupping," females of many species will move into specific nursery areas that are high in nutrients (food is available for the newborns) and that have a reduced potential of predation on the young. Although no cartilaginous fishes are known to provide parental care after birth, by placing their young in areas of high nutrients and low predation they ensure that a fairly high percentage of the young will survive. This is important, especially considering the low number of offspring they produce. Adolescent congeners are often excluded from the nursery grounds until they mature, perhaps to reduce competition between juvenile and adolescent animals or to remove a potential predator of the newborns, as adolescents of some species will prey on the juveniles of their own kind. These nursery areas are very specific for some species and the same individuals will return seasonally to these areas to give birth. Courtship and copulation also occur in these areas, although this is not always the case. Some species may move into bays, estuaries, or lagoons, others may move offshore, some deepwater forms may migrate up continental slopes and onto the continental shelf to give birth, and some bottom-dwelling species have their young migrate into the midwater zone.

Over the course of a year, changes in prey composition, water temperature, and salinity will variously affect the abundance and composition of the cartilaginous fish species residing in a particular habitat. Adults of some species may disappear for portions of the year, although their juveniles may remain within the confines of the nursery grounds until they reach adolescence. This is particularly evident in several California bays, which serve as important areas for mating and as nursery grounds. The seasonal return of some species to these bays follows a pattern known as "sequential migration." For example, the return of adult Sevengill Sharks to San Francisco and Humboldt Bays typically follows the arrival of adult female Bat Rays and Houndsharks, both of which subsequently give birth. Juvenile Bat Rays and houndsharks are important prey for young Sevengills and thus the appearance of these species in bays provides a significant food resource.

Age and Growth

Chondrichthyans, unlike bony fishes, have very few hard parts that can be used for estimating age. The age of a chondrichthyan is usually estimated by counting the number of calcified bands across the vertebral centra, in much the same way as the age of a tree is determined by counting the rings on its trunk. A band typically consists of one translucent (light) band and one opaque (dark) band (fig. 5). The deposition of opaque and translucent bands can sometimes be correlated with summer and winter seasons, respectively. In addition to vertebrae, the bands on the dorsal fin spines may also be an aid to estimating age. These bands are usually enhanced through various staining techniques, sectioning, or X-ray. However, the interval at which

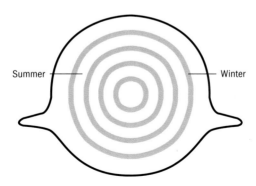

Figure 5. Cross-section of chondrichthyan vertebral centrum depicting light (summer) and dark (winter) bands.

these bands occur remains unknown for most species. Bands are generally assumed to be laid down once a year, but this has been validated for only a few species, such as the Leopard Shark and some of the smoothhound sharks. In the Basking Shark it appears that two bands are laid down every year. Furthermore, band deposition may not be correlated with age, as in the Pacific Angel Shark where it is associated with somatic growth. Pacific Angel Sharks are born with six or seven bands on their centra rather than the customary birth band observed in most species.

Size

Approximately 50 percent of all living sharks are less than 1 m in length—far smaller than the average size of a human—and 82 percent are less than 2 m long, with only about 18 percent of all shark species reaching a total length of more than 2 m. The average maximum length for living sharks is 1.5 m. Only 4 percent of all living shark species can be considered gigantic—sharks that regularly exceed 4 m and may grow up to 12 m or more in length. This includes the two largest living species, the Whale Shark and the Basking Shark, which may grow to lengths of 10 to 18 m. The Great White Shark, Sixgill Shark, Megamouth Shark, and Pacific Sleeper Sharks are other species known to exceed a maximum size in excess of 4.5 m. In contrast to these giants, some sharks are considered dwarfs or pygmies, reaching only 15 to 20 cm in length at maturity (e.g., the Pygmy Shark, which matures at a size of less than 20 cm). Most rays are less than 1 m in length at maturity, with fewer than 10 percent exceeding a length of 4 m or a width of 3 m. The largest rays include the Sawfishes, which can attain a length of 7 to 10 m, and the Mantas, which can have a disc width of 6 to 7 m. Chimaeras are small to moderately large, with most maturing at less than 1 m and only a few reaching over 1.5 m in length.

Food and Feeding Behavior

All living cartilaginous fishes are carnivorous predators consuming some type of animal protein for their diet. Although no known species are primarily herbivorous, some may inadvertently ingest algae while feeding. The size of prey items varies from minute crustaceans, ingested by some of the giant filter feeding sharks, to pinnipeds, cetaceans, and large oceanic fishes, consumed primarily by large sharks such as the Great White Shark or Tiger Shark. Some species are very selective feeders, focusing on bony fishes, crustaceans, or perhaps cephalopods, whereas others are extremely versatile and opportunistic predators. Most sharks have on occasion consumed some form of inorganic garbage, but by and large sharks tend to shy away from this type of food. The one exception is the Tiger Shark, which has a penchant for tasting unusual food items including leather, wood, coal, plastic bags, small barrels, cans, and other assorted junk related to human activity. Except for the Tiger Shark, which has often been called a garbage can with fins, the majority of sharks are not the blindly voracious predators of popular legend and generally feed on a very limited spectrum of mostly live prey. Cartilaginous fishes do not primarily scavenge, although some of the larger shark species will feed on carrion when available. Much of the misconception with regard to shark predation came from people watching them feed on refuse tossed over a ship or offal being discarded into the sea. As with any predator, sharks are unlikely to pass on an easy meal.

Sharks and rays, rather than being slow moving clumsy predators that feed more by chance, are actually extremely efficient hunters with a sophisticated predatory behavior pattern. The size and efficiency of the "hunter" relative to the size and behavior of its prey will greatly influence the way in which a shark or ray will attack. Depending on the species, variables such as size and shape, preferred habitat, life history stage, and season will be important factors in determining the size and type of prey it will consume. Studies on the foraging behavior of sharks and rays have shown that they possess a repertoire of strategies that they may employ when hunting specific prey items.

Perhaps the most interesting strategy by which sharks may feed, and one that greatly enhances the prey spectrum, is pack hunting (pl. 1). It was hypothesized over 30 years ago by Stewart Springer, a well-known shark biologist, that some of the tiny Lanternsharks (*Etmopterus* spp.) hunt in packs to subdue squids that were much larger in size than any one individual shark. Subsequent studies revealed that not only smaller sharks, but also larger species, including the Great White Shark, will hunt and feed cooperatively. These observations dispel the notion that all

Plate 1. Sevengill Sharks cooperatively hunting and feeding on a Harbor Seal (*Phoca vitulina*) in northern California.

sharks are lone marauders and indicate that many species are in fact quite social. Social hunting includes numerous advantages: sharks can (1) attack and subdue prey species that may be larger than any one individual predator, (2) forage over a wider area for food, (3) alert other sharks, through their behavior, to the presence of a food source, and (4) work cooperatively to herd numerous schooling prey species close together, thereby benefitting the whole group. Social facilitation in hunting prey items has been observed in Spiny Dogfish, Sevengill Sharks, Copper Sharks, Oceanic Whitetip Sharks, Blue Sharks, Hammerhead Sharks, and Bat Rays, among others. Likewise, rays will frequently forage in large groups searching for food, though less is known about their feeding behavior.

Other strategies that sharks and rays may incorporate in their hunting behavior include ambushing, burst speed, parasitizing, and filter feeding. Some sharks, and most rays, especially those that are not very mobile, catch their prey by ambushing it. They do this by lying in wait, like the Angel Shark, which lies partly buried in the sand or mud waiting for an unsuspecting prey item to swim by. Other sharks ambush their prey by stealth. With their neutrally buoyant livers sharks don't need to expend much energy to maintain their position in the water and thus are able to approach an unsuspecting prey item without creating any vibrations that the fish can detect. This is how apparently slow-swimming sharks are able to catch fast-swimming fishes such as tuna and salmon. Species such as the giant Pacific Sleeper Shark consume tunas in far greater quantities than it seems could have been scavenged. However, by approaching their prey in a very stealthy manner these sharks are able to get within striking distance before putting on a quick burst of speed. Several species, mainly oceanic forms, have tremendous speed and can catch their prey simply by running it down. The Mako and Salmon Sharks are excellent examples. Both are extremely active, fast-swimming species that commonly follow schools of tuna and salmon. The Great White Shark also is capable of catching fast-swimming species, as evidenced by a 5.5-m female shark caught off Anacapa Island that was found to have consumed two Blue Sharks, a Mako Shark, and two California Sea Lions.

Among these various feeding strategies are two forms of very specialized predation: parasitizing and filter feeding. The smallish Cookiecutter Shark (*Isistius* spp.) often attacks large Sword-fishes, Elephant Seals, and whales by taking a gouging bite out of them (pl. 2). This probably causes its victim more irritation than anything else. The giant filter-feeding species have each developed a unique way in which to feed. The Whale Shark may swim passively with its mouth open, but more often than not it will come up through a school of small fish or crustaceans at the surface, raise its head out of the water, open its mouth, and by slowly sinking back into the water draw in its food. The Basking Shark uses even less energy to catch its food. It simply swims along at a pace of about 3 to 5 km/hr with its giant mouth open and allows water to pass over its gills, which, like the baleen in whales, catches the food. The Megamouth Shark uses a similar passive means of filter feeding, only this species migrates up and down the water column following its prey. The Manta gracefully glides through the water with its mouth open, passively ingesting its prey.

Cartilaginous fishes may migrate considerable distances to follow their major food sources. This is particularly true for many of the pelagic species such as the Thresher, Salmon, and Blue Sharks, which feed heavily on schooling pelagic fishes and squids. The movement of these sharks in pursuit of their primary prey species is in large part related to the behavior of the prey species.

Many cartilaginous fishes have special adaptations for catching their food. The long upper lobe of a Thresher Shark caudal fin may be used to stun or kill its prey before eating it. The long saw-like rostrum of the sawfishes (Pristidae) is used in a rapid side-to-side motion to stun or kill prey as well as for defense. The long abdomen of the Frilled Shark along with its highly distensible mouth allows it to engulf prey items more than one-half its own length. Other adaptations are less obvious, such as the long flattened snout of the Goblin Shark, which may act as a sensor for detecting prey in its deep-sea environment.

Other aspects of chondrichthyan predation that are less well known include prey behavior, foraging success, and prey-related injuries to the predator. The Swell Shark tends to hunt at night when many of its prey items are resting or asleep. By foraging at night this apparently lethargic shark is able to capture relatively active species that it probably could not capture during the day. Basking Sharks will congregate in areas of high productivity,

Plate 2. Cookiecutter Sharks ambushing a school of Dolphinfishes
(*Coryphaena hippurus*).

usually from upwelling, where they will forage on certain plank-tonic species. The frequency of success of the predator is largely unknown, although scars on pinnipeds and cetaceans, which are more easily identified than scars left on a bony fish, may be one indication. Tooth scars left by sharks on Pacific Torpedo Rays would suggest that the electric voltage put out by these rays is sufficient to deter an attacker, yet this may be more the result of the size differential between the predator and its intended prey. A sufficiently larger predator would successfully subdue and consume the same Torpedo Ray. In addition to being used defensively, the electric organs of Torpedo Rays are used to hunt and subdue prey.

Another area of predatory behavior that is relatively unknown is prey-related injuries. The Torpedo Ray fending off a potential predator using its electric organs is one example. A swordfish bill was once found imbedded in a Mako Shark caught off Baja. The shark was alive when captured and the wound appeared to have healed. Spines from stingrays and various squaloid sharks are frequently found embedded in the head, mouth, and body cavity of sharks. It can be assumed that on occasion the predator may be killed by its intended prey during the attack.

Ecology

Cartilaginous fishes should be viewed as an integral and functioning part of any marine community, with each species occupying a distinct niche. The integral role these fishes play in the marine environment should not be underestimated, as in most instances they are among the apex predators. Yet despite their importance to the marine community, critical studies elucidating the precise role they play are seriously lacking. Much of the "research" conducted has not been beneficial, tending to focus either on the rare phenomenon of shark attack or on exploiting the public's imagination in the name of conservation or education. Historically cartilaginous fishes have often been lumped into a catchall category termed "sharks" or "rays" with little or no reference made to how many or what kinds of species are represented. This makes it difficult to study their ecological role.

Ecomorphology is the term used by ichthyologists to describe the study of the morphology and natural history of cartilaginous fishes in combination with their apparent life history.

Much can be inferred about a shark's life history by the design of its basic body form. The streamlined body form of fast-swimming oceanic sharks as exemplified by the Shortfin Mako and Blue Sharks is well designed for speed and agility. On the other hand the angel sharks, with their flattened, raylike body design, are well suited for a sandy bottom habitat. Many species of less active sharks, such as the Swell Shark, have a narrow, cylindrical body that is ideally suited for crawling over reefs and into cracks and crevices. The slightly compressed, flattened bodies of the batoids suit them well for cruising just off the bottom while searching for food. Although the general body form of most batoids is similar, they actually have a fairly complex suite of habitat requirements depending on the species and its preferred substrate type.

Ecosystems

Cartilaginous fishes may live on or near the ocean bottom, in what is called the benthic zone, or away from the bottom in the open sea, in what is known as the pelagic zone (fig. 6). Zone boundaries are crossed by numerous species of cartilaginous fishes and other marine animals, but because these zones are set up according to factors such as amount of light penetration,

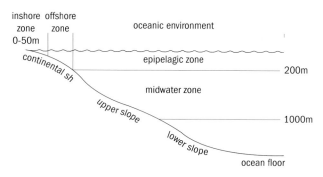

Figure 6. Generalized cross section of marine habitats along the California coast.

temperature, and the extent of the continental shelf and slope, they have considerable biological significance. Each of these two broadly classified zones, benthic and pelagic, is further subdivided into several distinct habitats, relative to the continental landmass, known as the continental shelf, continental slope, and oceanic habitats.

The continental shelf, which has a nearshore or coastal habitat that includes bays and estuaries, extends out to about 50 m in depth. Many of the cartilaginous fish species encountered in this zone are familiar to local anglers and others who spend time near the sea. Species that are common to the coastal environment and may be frequently encountered include the Sevengill Shark, Great White Shark, houndshark, Leopard Shark, Swell Shark, Horn Shark, Bat Ray, and Round Stingray. Beyond 50 m and out to a depth of approximately 200 m is the outer continental shelf. Species that are common to this outer shelf region include the Requiem Shark, Dogfish Shark, Thresher Shark, Hammerhead Shark, and Skate.

The cartilaginous fish fauna of a typical enclosed central or northern California bay environment can be differentiated both seasonally and spatially. Within a bay are a number of distinct habitats, from the deep central portion to its northern and southern reaches, where the bay fans out to form large expanses of mud flats that become exposed during spring tides. These mud flats are often bisected by deeper channels, formed by runoff from small creeks and rivers. The bay ecosystem may consist of several elasmobranch species, each inhabiting one or more of these distinct areas. Typical members of this environment include the Brown Smoothhound, Leopard, and Sevengill Sharks as well as the Bat Ray. Leopard Sharks and Bat Rays tend to forage in the shallows of the mud flats, rooting out clams, burrowing worms, and mud shrimp, whereas the Brown Smoothhound, inept at mud rooting, swims just off the bottom in the deeper channels in search of crustaceans. Also hunting along these channels and occasionally venturing onto the mud flats are more active, powerful predators such as the Great White Shark and Sevengill Shark, both of which feed on, among other things, houndsharks, Bat Rays, and pinnipeds.

Open coastal areas also contain a variety of habitats, ranging from rocky reefs with kelp forests to sandy or mud bottoms with little vertical relief. Each habitat supports a different community of elasmobranchs. Nearshore rocky reef communities often consist of Bat Rays, Leopard Sharks, Horn Sharks, and Swell Sharks, whereas sandy beaches are ideal for the uniquely flattened Pacific Angel Shark, which lies quietly on the bottom and is often seen by divers, and for Torpedo Rays, Guitarfishes, and Thornback Rays, which patrol the bottom, each in search of its favorite prey. Farther offshore, but still on the continental shelf, are several species of benthic and pelagic elasmobranchs, such as the Thresher Shark, Requiem Shark, Catshark, Dogfish Shark, Stingray, and Skate.

Beyond the shelf is the continental slope, which can be divided into an upper (200 to 1,000 m) and lower (greater than 1000 m) slope. The continental slope is rich in species, with the dogfish sharks, catsharks, and skates representing the major groups of benthic cartilaginous fishes. Some species, such as the Sixgill Shark, move from the continental slope onto the shelf to pup during spring, and it is not uncommon to catch newborn and subadult Sixgills in relatively shallow water. The bizarre-looking but poorly known chimaeras are most abundant in this environment.

The California coast is bisected by numerous submarine canyons, with the grandest being the Monterey Submarine Canyon in Monterey Bay. These submarine canyons have their own unique set of habitats. Several deepwater species, such as the Prickly Shark, will congregate seasonally at the head of some submarine canyons in relatively shallow water for reasons that are still unclear. These submarine canyons are also unique in that they offer researchers the opportunity to study many deeper dwelling species without having to travel a great distance from land.

The oceanic environment is that region in which species move about in the water column with no reference points such as the bottom or pinnacles. This is the area in which many of the fast-swimming fishes, such as tunas, mackerels, and billfishes, are found. The oceanic elasmobranch fauna in California waters is mostly dominated by warm-temperate and tropical species, with only a few cold-temperate water species represented. Oceanic sharks include a variety of near-surface and deepwater species that range in size from the tiny Pygmy Shark to the giant Whale Shark. Oceanic sharks are generally wide ranging, with their distribution dependent mainly on the temperature of the water. Many species, particularly at the edge of their distribution, will migrate seasonally with the movement of water masses. Although some species occasionally visit the continental shelf, they are usually found far out to sea. In the case of the Oceanic Whitetip

Shark, considered one of the species most aggressive toward humans, this is probably good news for swimmers. Some sharks are known to give birth on the continental shelf, but the young as they mature migrate outward to the open ocean. Still other species, such as juvenile Spiny Dogfish, spend their first few years in an oceanic environment before settling into a more benthic lifestyle as adults. There are no known pelagic chimaeras.

The oceanic environment can be further subdivided vertically into an epipelagic zone and a midwater zone. The epipelagic zone is that area ranging from the surface down to a depth of approximately 200 m. This area is influenced by the degree of penetration of sunlight. Many of the shark species that occur in this habitat, like their bony fish counterparts, are fast-swimming species such as the Mako Shark, Silky Shark, and Blue Shark. In addition, there are some weaker swimming species such as the Oceanic Whitetip Shark or the Crocodile Shark. The Pelagic Stingray is usually found in the uppermost 100 m in this environment. Below the epipelagic zone is the midwater zone, an area that does not receive sunlight. Sharks in this zone are generally weaker swimmers such as the gigantic Megamouth Shark. Many of the shark species that occur in this zone migrate vertically into the shallower epipelagic zone at night to feed. One species, the Pygmy Shark, has been recorded to occur in water over 9,000 m deep, perhaps migrating this tremendous distance daily in search of food. The Cookiecutter Shark has been recorded from near the surface to a depth of over 3,500 m. It would not be surprising if larger species, such as the Megamouth Shark, also were found at considerable depths in the ocean.

Fisheries

The earliest fishery for cartilaginous fishes in California, started by indigenous people living along the coast, concentrated on common coastal species. In the mid-1800s Chinese immigrants to California fished for various coastal sharks and rays, particularly in and around San Francisco Bay. Interestingly, several species first described by ichthyologists were collected from fish markets in the Chinese communities along the central coast. Prior to 1936 the meat from sharks and rays was generally discarded, except in the Asian communities, where it was consumed.

The average price of shark and ray meat at this time was $0.10 to $0.20 per pound, although the fins, which are used as soup stock, sold for $2.50 per pound or more. During the 1900s, shark and ray commercial fisheries in California surged and regressed, experiencing two major cycles.

The first major cycle developed between 1936 and 1938 with the gradual realization that the livers of Soupfin and Spiny Dogfish Sharks, rich in vitamin A, could be sold on a competitive basis with cod liver oil. The outbreak of World War II in 1939 resulted in the curtailment of cod liver oil production and exportation from Europe. The west coast shark population therefore represented a tremendous source of raw material. The market for shark liver oil to replace the nonavailable cod liver oil improved rapidly and our expanding industry was soon supplying vitamin oils to Europe. However, by the mid-1940s the huge potential of the Pacific coast Soupfin and Spiny Dogfish Sharks supply had been tapped, synthetic vitamins were developed, and the fishery finally collapsed. The second major fishery cycle started in the mid- to late 1970s after the movie *Jaws* raised public awareness of sharks. This fishery mainly revolved around sharks as food for human consumption. The main species taken by commercial fisheries were the Blue, Shortfin Mako, Common Thresher, Angel, Soupfin, and Leopard Sharks and the Leopard Shark by recreational anglers, with other species such as the Sevengill, Sixgill, Great White, and Salmon Sharks, several of the skates, and the Bat Ray as minor catches.

Typically sharks and rays are initially caught in large numbers, but due to their low fecundity, slow growth, and late age at maturity, their populations quickly collapse. Unlike bony fishes, which are highly fecund, cartilaginous fishes have a very low fecundity rate. Sharks in particular tend to be inquisitive by nature and somewhat fearless, which makes them particularly vulnerable to overfishing. Rays, especially the skates, are taken in considerable numbers as a by-catch to other commercially important species in bottom trawl fisheries worldwide. Failure to record these catch data results in a gross underestimation of the numbers of rays actually caught.

It is unfortunate that it took the overexploitation of our local shark populations to finally prompt fishery agencies to adopt stringent regulations. However, to develop effective management plans fishery agencies need to know more about the life history of the commercially important shark species, need to realize the limitations the characteristics of the typical chondrichthyan life history impose, and need to understand the consequences of pressure from heavy fishing on shark and ray populations. Otherwise those regulatory agencies responsible for managing these fisheries will continue to fail.

In terms of abundance the White-spotted Ratfish is the only common chimaera species, but it has never been fished in California waters.

Injuries from Cartilaginous Fishes

Between 1950 and 1999 there were 106 shark attacks, including 11 fatalities, reported along the California coast (table 3). This represented an average of 2.1 attacks per year during this 50-yr period. The Great White Shark was identified as the culprit in 85 of these attacks. Other species implicated include the Blue Shark (three), Leopard Shark (two), Mako Shark (one), Hammerhead Shark (one), Sevengill Shark (one), and Tiger Shark (one). In 12 attacks, the species was unconfirmed. In addition, Sevengills have been implicated in about a dozen attacks on dogs in and around San Francisco Bay. A bather in at least one attack was playing with his dog in shallow water when a 1.5- to 1.8-m-long "spotted" shark attacked. Although the shark was unidentified, the attack occurred inside San Francisco Bay in an area in which Sevengills are fairly abundant.

The majority of shark attacks occurred while people were engaged in recreational water activities ($n = 98$) versus a commercial activity ($n = 8$). Of those people attacked along the California coast 73 percent were either surfing (24), skin diving (17), swimming (16), or scuba diving (15). Over the past 10 to 15 years kayaking and windsurfing have become popular activities and, not coincidentally, there has been an increasing number of attacks on people engaged in these water sports.

Coastwide the majority of attacks have taken place north of Point Conception (74), with the Great White Shark being the primary culprit. Of these attacks 65 percent were in an area known as the red triangle, which stretches from Monterey Bay north to Bodega Bay and west to the Farallon Islands. Three locations, Tomales Point (seven), San Miguel Island (six), and the Farallon Islands (six), have the dubious honor of being the sites of most of the attacks (table 4).

TABLE 3. Number of Shark Attacks per Year along the California Coast between 1950 and 1999*

Year	Number of Attacks	Year	Number of Attacks	Year	Number of Attacks	Year	Number of Attacks	Year	Number of Attacks
1950	2	1960	2	1970	0	1980	1	1990	5
1951	0	1961	3	1971	1	1981	1	1991	4
1952	3	1962	2	1972	4	1982	4	1992	2
1953	0	1963	0	1973	0	1983	0	1993	5
1954	1	1964	1	1974	7	1984	3	1994	3
1955	4	1965	1	1975	5	1985	2	1995	4
1956	1	1966	1	1976	3	1986	1	1996	3
1957	2	1967	0	1977	1	1987	1	1997	1
1958	1	1968	1	1978	2	1988	2	1998	1
1959	7	1969	2	1979	1	1989	3	1999	2
Total	21		13		24		18		30

*Range: zero to seven attacks per year.

TABLE 4. Number of Shark Attacks by County and Location in California between 1950 and 1999

County	Total	Location	Total
Marin	12		
		Tomales Point	7
		Point Reyes	2
		Bird Rock	1
		Dillion Beach	1
		Stinson Beach	1
San Diego	12		
		La Jolla	5
		Imperial Beach	2
		Coronado	2
		Mission Beach	1
		Sunset Cliffs	1
		Oceanside	1
San Mateo	12		
		Pedro Beach	1
		Pigeon Point	3
		Point Purisima	1
		San Gregorio	1
		Ano Nuevo	2
		Tunitas	1
		Montara	1
		Linda Mar	1
		Franklin Point	1
Santa Barbara	11		
		San Miguel Island	6
		Santa Catalina Channel	1
		Point Conception	3
		Franklin Point	1
Monterey	9		
		Lover's Cove	1
		Pacific Grove	1
		Monterey	2
		Monastery Beach	2
		Point Lobos	1
		Point Sur	2

County	Total	Location	Total
Sonoma	8		
		Bodega Rock	2
		Salmon Creek Beach	2
		Sea Ranch	1
		Stillwater Cove	1
		Jenner	2
San Francisco	7		
		Baker's Beach	1
		Farallon Islands	6
Humboldt	7		
		Trinidad Bay	3
		Moonstone Beach	2
		Clam Beach	1
		Humboldt Bay	1
Los Angeles	7		
		Malibu	2
		Venice Beach	2
		San Pedro Channel	1
		Hermosa Beach	1
		Laguna Beach	1
Mendocino	6		
		Albion	1
		Bear Harbor	2
		Shelter Cove	2
		Westport	1
Santa Cruz	6		
		Santa Cruz	1
		Pajaro Dunes	1
		Davenport	3
		Waddell	1
San Luis Obispo	5		
		Pismo Beach	1
		Point Buchon	1
		Morro Bay	3
Ventura	2		
		Ventura	2
Del Norte	1		
		Klamath River Mouth	1
Orange	1		
		Seal Beach	1

The number of attacks is surprisingly small given the population of California (more than 30 million people and growing) and the fact that most people live near the ocean and/or participate in some form of marine activity. In the 1990s, with a rapidly growing ecotourism industry, the average number of attacks increased to 3.0 per year from an average of 1.8 per year in the 1980s. Along with California's growing human population, marine mammal populations are also on the rise, and although this has been good for the Great White Shark population, it may not be as good for beachgoers. Thus, increased interactions between humans and sharks are to be expected.

The number of attacks per year from 1950 to 1999 ranged from zero to seven with an average of 2.1 per year. Separating these 50 years into three categories based on number of attacks (zero to one, two to four, and five to seven) reveals that for 23 of these years there were zero or one attack for a total of 16 attacks. For the 22 years in which two to four attacks occurred there were a total of 61 attacks. Finally, looking at the five years in which the highest number of attacks took place there were 29 total attacks representing more than 25 percent of all attacks during this time. Why there are a higher number of attacks in some years compared with others is unknown, but this may be related in part to oceanographic conditions or human patterns of behavior.

Between 1950 and 1999 there were 26 years in which El Niño conditions persisted. During these years there were 54 attacks for an average of 2.1 attacks per year. Interestingly, although potentially dangerous warm-water species such as the Tiger Shark migrate into our area in El Niño years, there was no increase in shark attacks in these years as might have been expected. However, La Niña years did result in a slight increase, with an average of 2.5 attacks per year. The average number of attacks was 1.9 in years in which neither El Niño nor La Niña conditions prevailed.

Although oceanographic conditions are significant, patterns of human behavior and demographics may have a greater impact on the number of shark attacks. For example, it is well documented that young males tend to be attacked in higher proportions than other demographic groups. This being the case, it is interesting to note that during war years, with many young men away, the incidence of shark attacks decreased (1.25 attacks per year during the Korean War years of 1950 to 1953 and 0.9 per year between the Vietnam War years of 1963 and 1969. With the war in Vietnam over in 1973 the beaches were again filled with young men engaging in water sport activities, wet suits became increasing popular, and the number of shark attacks skyrocketed to 3.2 per year over the course of the remaining six years of the decade. The average number of attacks per year declined in the 1980s, most likely due to a combination of renewed interest in sharks as a commercial and recreational fishery, which caused the population of some sharks to decline, and severe El Niño and other oceanographic-related events in the early part of the decade. As the 1980s came to a close, recreational activities such as surfing, kayaking, and scuba diving became increasingly popular along with ecotourism. This trend continued throughout the 1990s, with 35 attacks, averaging about 3.0 per year, recorded between 1988 and 1999 (representing about one-third of all shark attacks that occurred between 1950 and 1999). Interestingly, in 1997 and 1998 (El Niño years), with the first of the "Baby Boomer" generation having turned 50, the number of attacks declined slightly, with only a slight increase in 1999 (a La Niña year!). It may be that as the population ages and the "Boomers" settle into a midlife lifestyle, the number of attacks will decline. Only time will tell!

Cartilaginous fish species other than sharks can inflict serious injuries on humans. Stingrays with their sharp, barbed, stinging spine can inflict a painful sting on people wading in the water. Seal Beach in southern California is notorious for having bathers stung by round stingrays, which use this popular beach area as a nursery ground (table 5). In 1963 over 500 people were stung by these rays in a 10-week period. In 2000, 385 people were stung on the northern end of Seal Beach, an area referred to as "Ray Bay." Because so many people have been stung by Round Stingrays in this area, local officials have started to clip the spines from hundreds of these rays to reduce the chances of bathers being stung. Removal of the spines does not injure the ray as they replace their spines annually. The spines of the Pelagic Stingray have been implicated in at least two fatalities. The dorsal fin spines of sharks and chimaeras can also inflict painful, although not fatal, wounds if mishandled. The Pacific Torpedo Ray can discharge 45 volts or more of electricity, enough to knock down a grown adult.

TABLE 5. Stingray Injuries at Seal Beach in Southern California between 1993 and 2001*

Year	Stingray Injuries
2001	299
2000	385
1999	290
1998	185
1997	132
1996	167
1995	209
1994	378
1993	31

*Injury statistics provided by Seal Beach lifeguards.

How to Use This Book

California's cartilaginous fish fauna should be no more difficult to identify than many of the bony fishes, birds, or marine mammals that occur along the coast. The common coastal species in particular are easily distinguishable from each other. Some of the less common to rare deepwater catsharks and skates, or those species that occasionally visit our area, may be a bit more difficult to identify. Most of the sharks in our area can be identified by focusing on particular characteristics, such as general body shape, number of gills, the presence or absence of fin spines, the position of the fins, body coloration, and tooth shape. Distinguishing batoid characteristics include the disc and tail shape, body color, the size and shape of the dermal denticles or enlarged thorns on the dorsal surface of the disc, and the presence or absence of tail spines. The chimaeras that occur off our coast can be distinguished from each other by their coloration, snout shape, and the presence or absence of an anal fin.

Experience is ultimately the best means by which you can quickly identify a species in the field. Keep in mind that a species not previously reported in our local waters may stray into our area. During extreme El Niño years a species commonly found in Mexican waters might migrate northward out of its usual range following the warmer water masses into our area. This is especially true for several of the requiem or hammerhead sharks, which can prove problematic even for the experienced ichthyologist. If you happen to collect or observe a particularly difficult-to-identify specimen, be sure to note where and when it was caught as it may be new to California. If the specimen is too large to keep be sure to take a good photograph of it in side view and save the teeth, jaws, and/or spines. Most local natural history museums, public aquariums, and universities have ichthyologists on staff who will gladly identify difficult specimens. A partial list of these institutions is provided at the end of this book.

Species Account

Under each species account is a color illustration of that particular species with a line illustration of the underside of the head and teeth. Following the illustrations is a description that can be used to identify the species, and pertinent information on its habitat and range, natural history, human interactions, nomenclature, and references. The species descriptions have been kept fairly simple, but it is advisable to consult the glossary and to examine the terminology illustrations (Figs. 7–9) to become familiar with the terminology used to identify cartilaginous fishes.

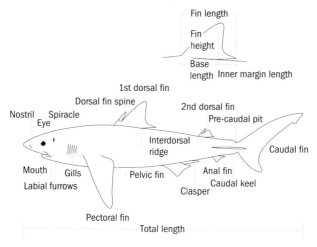

Figure 7. Shark terminology.

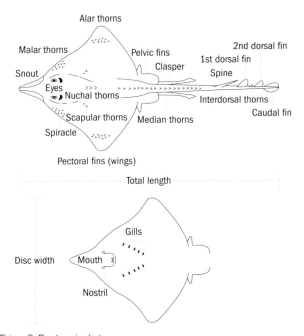

Figure 8. Ray terminology.

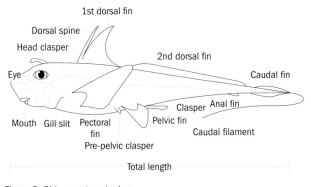

Figure 9. Chimaera terminology.

Illustrations

Color illustrations for all chondrichthyan species from off the California coast are based on the most "common" color scheme for each. In species whose color may vary regionally the most common local variant is illustrated. Juvenile coloration in newborns may vary considerably from that of the adult, although some species are very rare and have poorly described color features. This is particularly true for the skates and other rarely seen deepwater species. Every effort has been made to describe the color of living specimens where possible or to follow closely the description in the literature. In those instances in which color variants may differ dramatically these differences are described.

A line drawing of the underside of the snout and a representative upper and lower tooth is also included for most species. The teeth on many species may vary by position, and in some exhibit sexual dimorphism, but every attempt has been made to provide a drawing that should assist in identifying the species. Additional tooth descriptions have been provided under the species account.

Common and Scientific Names

The common names used here are taken from the American Fisheries Society (1991) publication, *Common and Scientific Names of Fishes from the United States and Canada,* or from the local vernacular used by fishermen, researchers, and others in referring to a particular species. Because the common name can change somewhat over the years, the most recent common name is used for each species. The scientific names used follow Compagno (1999) for the sharks and rays and Didier (1995) for the chimaeras.

Description

In this subsection a brief descriptive account of each species emphasizing key features such as body, snout, eye, mouth, fin, and tooth shape; number of gills; the presence or absence of dorsal and anal fins; the presence or absence of fin spines; and the relative position of fins to their approximate origin and/or insertion where appropriate is given. Ranges in the total number of (upper and lower) teeth, vertebra, and spiral valve counts are included, and although they may be of more use to ichthyologists, the amateur naturalist shouldn't be intimidated from using them to identify species. The counts used in this book come from several sources including my own research, unpublished data generously provided by colleagues, and literature accounts including, but not exclusively, those by Compagno (1988, 1990), Ebert (1990), Garrick (1982), Gilbert (1967), Nishida (1990), Notarbartolo-di-Sciara (1987), and Springer and Garrick (1964). Counts were taken for eastern North Pacific specimens except where little or no local information was available. An asterisk (*) is used to denote counts taken for specimens outside the eastern North Pacific region.

Habitat and Range

The geographic range in our waters is given as well as the range throughout the eastern North Pacific and general geographic distribution if applicable.

Natural History

The biological information for this subsection is based on original data I have collected over my many years of studying California's cartilaginous fish fauna, as well as data generously provided by colleagues, and from the many references cited in this book. Included in this subsection is information on species' mode of reproduction, size at maturity, size at birth, maximum size, migratory patterns, age at maturity, longevity, growth rate, diet, and foraging behavior as well as predators that feed on cartilaginous fishes.

Human Interactions

This subsection discusses the relationship between humans and cartilaginous fishes including commercial and recreational fisheries and injuries to humans caused by some of these fishes.

Nomenclature

The derivation of the Greek or Latin scientific name is included in this subsection as well as other common names that have been used in the literature. Common names from areas outside California waters are not included. Also in this subsection is a brief synonymy in instances in which taxonomic confusion has existed with regard to a particular species. The synonymy for each species is not complete, but includes those names frequently cited in earlier publications on the California fauna.

References

This subsection includes pertinent or significant literature based on studies from California or the eastern North Pacific. Additional references that were consulted for information on California's cartilaginous fish fauna, but that are not listed under each species account, include Castro (1983), Compagno (1984, 1988), Eschmeyer et al. (1983), Jordan and Evermann (1896), Roedel and Ripley (1950), Kato et al. (1967), Hart (1973), Miller and Lea (1972), Starks (1917, 1918), and Walford (1935).

Taxanomical Keys

A series of keys has been provided so that the reader can narrow down the list of options in identifying a particular species. At the end of this introduction is a key to the orders that will tell you whether the specimen is a shark, a ray, or a chimaera. By considering each pair of options and choosing the one that most accurately describes your specimen—regarding number of gill openings, presence or absence of fins, shape of the body, and so on—you will by a process of elimination arrive at the correct order.

Next, turn to the listing for that order (using the color coding in the table of contents) and you will mostly likely find a key to the families that comprise it. (There is no family key for the orders Squatiniformes, Heterodontiformes, and Orectolobiformes because they are monotypic, which means they are represented by only one family and one species in California waters.) As with the key to the orders, just work your way through the paired options to determine the family that your specimen is in.

From there, turn to the appropriate family—these are arranged phenotypically rather than alphabetically, so you may need to hunt through the order section a bit. In some cases, such as that of the frilled shark (Chylamydoselachidae), a single species is found; in others, such as the requiem sharks (Carcharhinidae), several genera are represented. Once you feel that you have keyed out the correct species, check to see if the description and illustration accurately describe your specimen. Remember that the color pattern shown is just one of several possible variants, and be sure to compare descriptions and illustrations of species in the same genus.

KEY TO CHONDRICHTHYAN ORDERS FOUND IN CALIFORNIA WATERS

1a   One paired gill opening...... Ratfishes (**Chimaeriformes**)
1b   Five to seven paired gill openings........................2
      2a   Anal fin absent 3
      2b   Anal fin present.................................5
3a   Body flat, raylike.......................................4
3b   Body not raylike.......... Dogfish Sharks (**Squaliformes**)
      4a   Mouth ventral, pectoral fins attached to head........
                      .... Rays or Batoids (**Rajiformes**)
      4b   Mouth terminal, pectoral fins not attached to head...
                      .... Angel Sharks (**Squatiniformes**)
5a   One dorsal fin, six or seven paired gill openings...........
               .... Cow and Frilled Sharks (**Hexanchiformes**)
5b   Two dorsal fins, five paired gill openings 6
      6a   Fin spines present................................
                  .... Horn Sharks (**Heterodontiformes**)
      6b   Fin spines absent...............................7
7a   Mouth in front of eyes...................................
                     .... Carpet Sharks (**Orectolobiformes**)
7b   Mouth behind eyes 8
      8a   Nictitating eyelids................................
                  .... Ground Sharks (**Carcharhiniformes**)
      8b   No nictitating eyelids.............................
                  .... Mackerel Sharks (**Lamniformes**)

SPECIES ACCOUNTS

COW AND FRILLED SHARKS (HEXANCHIFORMES)

The order of cow and frilled sharks comprises two families, four genera, and five to six species of moderate-sized to very large sharks; both families, three genera, and three species occur in California waters. These sharks are unique in that they are the only species to combine six or seven paired gill slits, a single dorsal fin, and an anal fin. The hexanchoids are usually considered one of the most primitive groups of modern-day sharks. They are a very poorly known group and one that has frequently been overlooked in favor of more diverse and better known shark species. Although lacking any specialized body features—for example, fin spines—the hexanchoids have adapted themselves to an extremely broad range of habitats. This fact is especially interesting because, given the low number of species in this group, their wide-ranging distribution compares favorably with the more species-rich Carcharhiniformes.

1a Body eellike, six paired gill slits with the first extending across the throat . . . . frilled sharks (Chlamydoselachidae)

1b Body fusiform, six or seven paired gill slits without the first extending across the throat . . . . cow sharks (Hexanchidae)

Frilled Sharks (Chlamydoselachidae)

The frilled shark family is represented by at least two species, one of which occurs off California. These distinctive, eellike, medium-sized sharks grow to between 1 and 2 m in length. Frilled Sharks are a deep-living species that are usually caught on or near the bottom in waters 120 to 1,450 m deep, although they are known occasionally to make excursions into the midwater zone. These sharks have one of the longest gestation periods—up to 3.5 years—among living elasmobranchs. Their diet consists mostly of cephalopods, other sharks, and bony fishes.

FRILLED SHARK *Chlamydoselachus anguineus*

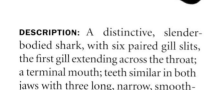

DESCRIPTION: A distinctive, slender-bodied shark, with six paired gill slits, the first gill extending across the throat; a terminal mouth; teeth similar in both jaws with three long, narrow, smooth-edged cusps, each being separated by an intermediate cusplet; a single dorsal fin; a very large anal fin; and a caudal fin without a subterminal lobe or notch. Coloration is a uniform dark brown or gray. Tooth count: 25/23. Vertebral count: 146–171.* Spiral valve count: 26–49.*

HABITAT AND RANGE: Very little is known about the habitat of this uncommon deepwater species. Most specimens have been taken over soft bottoms or in midwater trawls and gill nets. They are known to vertically migrate into the water column, often ascending more than 1,500 m to the surface. These vertical migrations occur mainly at night and appear to be related to feeding. The worldwide distribution of this species appears to coincide with regions of high biological activity, such as upwelling zones. Hydrographic data taken at the time of capture for one specimen revealed that it was in a low-oxygen, high-nutrient zone.

The only record of this species from California waters came from a specimen taken in a drift gill net at a depth of 20 m in water that was over 1,500 m deep. The specimen was caught about 22 miles southwest of Point Arguello, Santa Barbara County. There have been reports of additional specimens taken off the Channel Islands, and in particular San Clemente Island, but these have not been confirmed. However, in light of other spectacular catches—which include the Megamouth and Goblin Sharks—around the Channel Islands, additional specimens might eventually be caught. Elsewhere, the Frilled Shark has a wide but scattered distribution, with most specimens having been reported in Japan and the eastern North Atlantic.

NATURAL HISTORY: Viviparous, without a yolk sac placenta, with litters of two to 15, six being the average. Gestation has variously been reported as two to 3.5 years. The young are 50 to 60 cm at birth. The smallest free-swimming specimen recorded measured 54 cm. Males mature at approximately 1 m and grow to at least 1.65 m, whereas females mature between 1.4 and 1.5 m and grow to at least 1.96 m.

Frilled Sharks are known to feed on cephalopods, other sharks, and bony fishes, but very little else is known about their diet. The elongated abdomen, inwardly projecting tooth arrangement, terminal mouth, and highly distensible jaws give these sharks a specialized body arrangement reminiscent of two teleost families: gulper eels (family Saccopharyngidae) and viperfish (family Chauliodontidae). Frilled Sharks are capable of swallowing prey items over one-half of their own body length, much the way snakes are able to engulf large prey items. The unique body shape of the Frilled Shark is ideal for grasping and engulfing prey, and its ability to forage in both demersal and midwater habitats suggests a highly evolved lifestyle.

HUMAN INTERACTIONS: These sharks are too rare in California waters to be of importance to fisheries.

NOMENCLATURE: *Chlamydoselachus anguineus* (Garman, 1884). The generic name is derived from the Greek *chlamy* for frill, the Greek *selachus* for shark, and the Latin *anguineus* for snakelike. The common name refers to the frilled appearance of the first gill slit.

REFERENCES: Noble (1948); Ebert (1990).

Sixgill and Sevengill Sharks (Hexanchidae)

The sixgill and sevengill sharks, or cowsharks, are a small family consisting of three genera and four living species, two of which are represented in California waters. These sharks have six or seven paired gill slits, a single dorsal fin, an anal fin, and an elongated caudal fin. Ranging in length from 1.4 m to nearly 5 m, these sharks are distributed widely, from cold-temperate to tropical seas. Habitat also ranges widely, from nearshore coastal waters, including bays and estuaries, to depths of over 2,500 m. The reproductive mode is viviparous without a yolk sac placenta and with litters of 13 to 108, depending on the species. These sharks feed on a wide variety of prey items, including cephalopods, crustaceans, other sharks, rays, bony fishes, and marine mammals.

1a Six paired gill slits . . . .
. . . . Sixgill Shark (*Hexanchus griseus*)
1b Seven paired gill slits . . . .
. . . . Sevengill Shark (*Notorynchus cepedianus*)

SIXGILL SHARK   *Hexanchus griseus*

DESCRIPTION: A giant, heavy-bodied shark with a broad, blunt snout; moderately large eyes that glow a bright green in life; six paired gill slits; a large, broadly rounded mouth; lower anterior-most teeth comb-shaped; and a single dor-

sal fin. The dorsal coloration varies from a dark blackish gray to a dark chocolate brown, with lighter shades of each color extreme. The ventral region is lighter, ranging from pale white to dirty gray. A visible light streak along the lateral line extends from midway up the caudal fin to about the pectoral fin base. Coloration of newborns is particularly brilliant, with the anterior edges of the fins whitish, including the dorsal fin apex and caudal fin dorsal margin. Tooth count: 26–46/20–36 (including smaller pebblelike posterior teeth). Vertebral count: 128–145. Spiral valve count: 35–39.

HABITAT AND RANGE: A deepwater species usually found along the continental shelf and upper slopes down to at least 2,500 m, although they may migrate up to several hundred meters off the bottom, occasionally coming to the surface. Juveniles are often caught close to shore, in bays including San Francisco's, whereas adults are normally taken in deeper water. These sharks are associated with areas of upwelling and high biological productivity. Hydrographic data variously taken in areas where Sixgills have been observed at depths of several hundred meters reveal a bottom temperature of 43 to 50 degrees F for waters with high nutrient levels.

Sixgills occur from the outer coast of Baja California northward to Alaska. Elsewhere, this is perhaps the most wide-ranging shark species other than the Blue Shark, with a circumglobal distribution from cold-temperate regions to the tropics. It may even be a polar species, although there are no confirmed records from this region.

NATURAL HISTORY: Viviparous, without a yolk sac placenta, with large litters of 47–108. Males mature at about 3.1 m and females at about 4.2 m. This is a large shark, with males reaching at least 3.5 m and females at least 4.8 m. The young are 61 to 73 cm at birth. Females move onto the continental shelf during the spring to give birth after a two-year reproductive cycle. Juveniles usually remain on the shelf and uppermost continental slopes until they reach adolescence, when they move into deeper water. Adult males typically remain in deep water, where courtship and mating take place.

Little is known about their growth rate, although juveniles held in captivity will grow quite rapidly, nearly doubling in size in their first year of life.

The Sixgill Shark is a powerful predator that feeds on a wide variety of prey species, including other sharks; rays; chimaeras; teleosts such as herring, hake, mackerel, halibut; and marine mammals such as seals and small cetaceans. Larger Sixgills will actively forage on quite large prey items, including Swordfishes, marlins, dolphinfishes, seals, and dolphins. They have been observed consuming whales as carrion. Juveniles held in captivity have a voracious appetite.

HUMAN INTERACTIONS: There is no directed fishery for Sixgills in California, although when caught incidentally they are readily sold at market. Elsewhere, this shark is utilized for its meat and liver oil.

The Sixgill is a potentially dangerous species, but because of its deepwater habitat it rarely comes into contact with divers and swimmers and consequently has never been implicated in an attack. In recent years, an ecotourism industry for recreational scuba divers to observe these sharks has sprung up in the Pacific Northwest. Care should be taken by anyone coming into contact with this shark as it can inflict severe injury if mishandled.

NOMENCLATURE: *Hexanchus griseus* (Bonnaterre, 1788). The generic name is derived from the Latin *hex* for six and *anchus* from *branchus* for gill. The species name *griseus* means gray.

The common name, Sixgill Shark, comes from the presence of six paired gill slits. This shark has been variously referred to as the Sixgill Cowshark, the Mud Shark, and the Shovelnose Shark.

Eastern North Pacific Sixgill Sharks were described as a distinct species, *H. corinus,* by Jordan and Gilbert (1880a) based chiefly on the number of large upper anterior lateral teeth (six versus eight or nine). However, subsequent research determined that the number of upper anterior lateral teeth increases with growth, thus invalidating the principle characteristic distinguishing *H. corinus* from *H. griseus.*

REFERENCES: Ebert (1986a,b, 1990, 1994); Jordan and Gilbert (1880a).

SEVENGILL SHARK   *Notorynchus cepedianus*

DESCRIPTION: The only shark in our area with seven paired gill slits; it has a broad head, blunt snout, relatively small eyes, comb-shaped lower anterior teeth, and a single dorsal fin. The dorsal background color varies from a pale silvery gray to reddish brown for specimens from Humboldt Bay to an olive brown to muddy gray color for Sevengills from San Francisco Bay. These distinctive background colors may represent distinct populations of Sevengills in northern California. The presence of irregular black spots dorsally, shading to a lighter coloration ventrally, is consistent in all specimens.

Sevengills in captivity lose their spots within several weeks of placement in display tanks. It is believed that Sevengills may be able to alter their coloration slightly. The loss of spotting in aquarium display animals may be in response to the artificial lighting.

White spots occasionally found on the back and sides are attributable to a fungus. Juvenile coloration is quite striking with the trailing fin edges, including the dorsal fin apex, being white. Albino and piebald specimens have been reported in California waters. Tooth count: 21–42/20–37 (including smaller posterior teeth). Vertebral count: 123–130. Spiral valve count: 14–17.

HABITAT AND RANGE: Sevengills are a common coastal inhabitant, appearing close inshore, in bays and estuaries, and along the continental shelf out to a depth of at least 136 m. They seem to be most abundant in areas in which the water temperature is 54 to 65 degrees F. These sharks are associated with areas of upwelling and high biological productivity. They tend to prefer rocky reef habitats where kelp beds thrive, although Sevengills are also commonly caught over sandy and mud bottoms. Although relatively common at certain times of the year in Humboldt and San Francisco Bays, very little is known about their movement patterns along the open coast.

Sevengills occur along the entire California coast, ranging from southeast Alaska to southern Baja California. A population occurs in the northern Gulf of California, but it is unknown whether this is continuous with the coastal population. Elsewhere, Sevengills occur in most temperate seas, including the southeastern and western Pacific, southwestern Indian Ocean, and southern Atlantic Ocean.

NATURAL HISTORY: Viviparous, without a yolk sac placenta, with litters of up to 82 young. Males mature at 1.5 to 1.8 m and grow to at least 2.5 m, whereas females mature at 2.2 to 2.5 m and grow to 3 m. The young are 35 to 45 cm at birth. Adult females have a 2-yr reproductive cycle, with Humboldt and San Francisco Bays serving as important birthing and nursery grounds. Newborn and juvenile Sevengills tend to remain in the vicinity of these nursery grounds until they begin adolescence, at which time they leave the area.

Juvenile Sevengills grow quite rapidly during the first two years of life, more than doubling their birth length. This rapid growth by juveniles in the nursery ground enhances their chance at survival as a Sevengill over 70 cm has far fewer predators than a newborn half its size. In contrast to the rapid growth of juveniles, maturing Sevengills have a much slower growth rate.

Sevengills are powerful predators whose diet includes teleosts, other elasmobranchs, and marine mammals, including seals and dolphins. Other than the Great White Shark they appear to be the dominant elasmobranch predator in the nearshore marine environment. Sevengills have been observed to employ a variety of foraging strategies. As a solitary hunter it uses stealth to ambush smaller prey items, but while hunting larger prey these sharks will hunt cooperatively in packs to subdue seals, dolphins, other large sharks, and rays. A Sevengill was once observed to beach itself from the force of an attack on a Leopard Shark. It then simply wiggled itself back into the water and swam off, its prey in mouth. Great White Sharks are one of the few sharks known to prey on adult Sevengill Sharks, and they have been observed to attack them on occasion. Conversely, Sevengills were once observed to attack a captive Great White Shark on display in a public aquarium.

HUMAN INTERACTIONS: Sevengill Sharks were one of the three main shark species (the others being the Soupfin and Spiny Dogfish) caught for their liver oil during the shark fishery boom years of the 1930s and 1940s. This was due in part to their seasonal abundance in Humboldt and San Francisco Bays. Even after this fishery industry collapsed, these sharks were taken in considerable numbers during fishing competitions in San Francisco Bay during the 1950s and 1960s. The popularity of the *Jaws* movies renewed interest in shark fishing in the 1970s and numerous commercial recreational boat operators in the San Francisco Bay Area chartered their boats specifically targeting Sevengills and Soupfins. Prior to the boom years of shark fishing these sharks were fished for their meat and fins going back to the mid-1800s by Chinese immigrants living in the area.

Primarily caught by recreational anglers, Sevengills are also taken incidentally by commercial fisheries seeking other species. Although not as popular as other shark species, the meat is of good quality and is often sold simply as "shark," thus making it difficult to quantify landings for this species.

Sevengills are a potentially dangerous species implicated in at least one attack on a human in California. In addition, there are more than a dozen records of this shark attacking dogs playing in water less than 1 m deep. Care should be taken by anglers in handling this shark as it can inflict severe injury if mishandled.

NOMENCLATURE: *Notorynchus cepedianus* (Peron, 1807). The scientific name *Notorynchus* refers to the spotting on the back of the Sevengill and *cepedianus* is in honor of the eighteenth-century naturalist B. G. E. Lacepede. The common name, Sevengill Shark, comes from the presence of the seven paired gill slits. Other local names ascribed to this shark include Cowshark, Sevengill Cowshark, and Broadnose Cowshark.

The Sevengill was first described by Peron (1807) as *Squalus cepedianus* based on a specimen from Tasmania, Australia. The species was later described regionally as being distinct from the Australian form by various authors, including Ayres (1855a), who erected the genus, *Notorynchus,* for which he described a new Sevengill Species, *N. maculatus,* from California waters. Other local researchers changed the genus to *Heptranchus* (Jordan and Gilbert 1880a) and later to *Heptanchus* (Daniel 1934), but *Notorynchus,* having seniority over the other generic names, was eventually recognized as the valid genus name. The species name, *maculatus,* as well as the other regional scientific names were generally retained until Ebert (1990), in a comprehensive study on the systematics of the order Hexanchiformes, concluded that the genus was monotypic with *N. cepedianus,* being the only valid species.

REFERENCES: Ebert (1985, 1986b, 1989, 1991); Van Dykhuizen and Mollet (1992).

DOGFISH SHARKS (SQUALIFORMES)

The dogfish sharks are the second largest shark order with seven families, 23 genera, and at least 100 species described worldwide. The number of species described is expected to increase with continued exploration of the deep sea. Five families representing five or six species occur off the coast of California. The dogfishes are a morphologically diverse group that contains some of the smallest as well as some of the largest known shark species. This diverse order is characterized by two small to moderately large dorsal fins usually preceded by a spine; no anal fin; a short to moderately long preoral snout; eyes without a nictitating membrane; a small to large spiracle; five paired gill slits, all of which are anterior to the pectoral fin origin; a broadly arched or short transverse subterminal mouth; moderately differentiated teeth, with many species having powerful jaws for cutting; and light-producing organs in several species. In cold-temperate and arctic seas they may occur in shallow water, but in warm-temperate and tropical waters they are replaced by more advanced requiem and hammerhead sharks. In the tropics some are pelagic, but none regularly occur inshore. The group includes the only known truly polar shark species. Dogfish sharks are found inshore to a depth of over 6,000 m. The taxonomic status of this order is currently in flux, with several authors questioning the arrangement of this taxon. Here I follow Compagno (1999).

1a Dorsal fins without fin spines 2
1b Dorsal fins with fin spines 4
    2a First dorsal fin originating behind pelvic fins . . . .
    . . . . bramble sharks (Echinorhinidae)
    2b First dorsal fin originating in front of pelvic fins 3
3a First dorsal fin originating closer to pelvic fins; second dorsal fin base about four times the length of the first dorsal fin base; maximum length 27 cm. . . . .
. . . . kitefin sharks (Dalatiidae)
3b First dorsal fin originating closer to pectoral fins; second dorsal fin base about equal to or shorter than first dorsal fin base; maximum length approximately 440 cm . . . .
. . . . sleeper sharks (Somniosidae)

4a Teeth with prominent central cusp flanked by one or two smaller cusplets . . . . lanternsharks (Etmopteridae)

4b Teeth oblique and bladelike with a single smooth-edged cusp . . . . dogfish sharks (Squalidae)

Bramble Sharks (Echinorhinidae)

The bramble sharks are a small group consisting of one genus and two species, one of which occurs off the coast of California. These are moderately large sharks with a stout, cylindrical body; a broadly arched mouth; small spiracles; no dorsal fin spines; and the first dorsal fin located over the pelvic fins. Their most noticeable feature is large tacklike dermal denticles covering their body. These are mainly deepwater sharks found on the outer continental shelves and upper slopes, occurring at a depth of at least 1,200 m, although they are occasionally taken in relatively shallow water, off beaches and in bays.

PRICKLY SHARK   *Echinorhinus cookei*

DESCRIPTION: A moderately large, stout, flabby body covered with very large, conspicuous, thornlike dermal denticles irregularly scattered over its body and fins; a first dorsal fin that originates behind the pelvic fin; and bladelike teeth, with a single smooth-edged cusp and one to 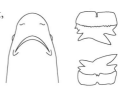 three smaller cusplets. They are a uniform brown to slate gray or black, with lighter coloring around the mouth and ventral surface of the snout. Their posterior fin margins are black. Tooth count: 21–25/20–27. Vertebral count: 88–92. Spiral valve count: 8–13.

HABITAT AND RANGE: A sluggish bottom-dwelling shark of the continental shelf and upper slopes found in soft mud or sandy bottoms. Although predominately a deepwater species found at a depth between 100 and 650 m, this shark has been known to move inshore into relatively shallow water. A Prickly Shark was once taken in 4 m of water inside the Moss Landing Harbor in Monterey Bay. Prickly Sharks can tolerate areas of low oxygen levels such as those found in deep ocean basins, which are unsuitable areas for sharks with higher oxygen requirements. Prickly Sharks appear to make seasonal inshore migrations to the heads of submarine canyons and onto the continental shelf and upper continental slopes. They prefer deep, cool waters with a temperature of 45 to 50 degrees F.

Prickly Sharks periodically congregate in groups of 30 or more at the head of the Monterey Submarine Canyon in water less than 35 m deep. Observations indicate that they tend to orient themselves close to the bottom or near the walls of submarine canyons.

Prickly Sharks occur from Moolach Beach, Oregon, to the Gulf of California and are sporadically distributed around the rest of the Pacific Ocean.

NATURAL HISTORY: Viviparous, without a yolk sac placenta. Large litters of up to 114 make them one of the most fecund of all shark species. Males mature at about 2.4 m and females at approximately 3 m. The maximum length is about 4 m. The size at birth is 35 to 45 cm.

Prickly Sharks feed mainly on fishes, including mackerel, hake, flatfish, rockcod, and other sharks including newborn Sixgills and Spiny Dogfish, as well as cephalopods and crustaceans. The morphology of the jaw and buccal cavity suggests that these sharks feed by suctioning to inhale their prey.

Sixgill Sharks are known to prey on young Prickly Sharks, but the adults probably have very few predators.

HUMAN INTERACTIONS: Prickley Sharks are occasionally caught in bottom trawls and on long-lines but are of no commercial importance. The meat is soft and of poor quality.

Despite their large size Prickly Sharks are not considered dangerous to people. They do not appear to be disturbed by the presence of divers and typically flee quite rapidly when approached. However, common sense should always be exercised when scuba diving near large sharks, even seemingly harmless species.

NOMENCLATURE: *Echinorhinus cookei* (Pietschmann, 1928). The generic name is derived from the Greek *echino,* meaning prickly, and *rhine,* meaning filelike or rough. Victor Pietschmann named the species *cookei* after a colleague, C. Montague Cooke Jr., in appreciation for his helpful assistance. The common name refers to the prickly thornlike denticles covering this shark. This shark has been referred to locally as the Bramble Shark.

The Echinorhinidae is sometimes placed in the family Squalidae, although morphological studies on their skeletal structure suggest that they are very distinct and should be retained in their own family, if not their own order. It was thought for some time that this family and genus were represented by a single wide-ranging species, the Bramble Shark (*E. brucus*). The Prickly Shark (*E. cookei*) was described in 1928 as being distinctly different based on a specimen caught off the Hawaiian Islands. Early California records of this shark referred to it as *E. brucus,* regarding *E. cookei* as a junior synonym. There are no records of *E. brucus* from the eastern Pacific Ocean.

REFERENCES: Barry and Maher (2000); Crane and Heine (1992); Varoujean (1972).

Dogfish Sharks (Squalidae)

The dogfish sharks were at one time the second largest family of sharks—the catsharks, Scyliorhinidae, being first. Current taxonomic revision of the family has divided the Squalidae into four additional families, three of which occur in the California area. The Squalidae are composed of two genera and at least 11 valid species, one of which occurs in the California area. These are small to medium-sized sharks, with a stout to slender body; a fifth gill opening similar in size to the first four; a short and transverse mouth; absent photophores; the first dorsal fin originating closer to the pectoral bases than to the pelvic bases; and each dorsal fin preceded by a fin spine. These sharks are wide ranging and generally inhabit offshore coastal waters on continental shelves and upper slopes. They usually occur in deeper water in the tropics. Worldwide some members of this family have historically been an important fishery.

SPINY DOGFISH

Squalus acanthias

DESCRIPTION: A slender-bodied dogfish with a relatively long, pointed snout; a first dorsal fin that originates just behind the free rear-tip of the pectoral fins; and a prominent fin spine preceding each dorsal fin. The teeth are oblique, bladelike, and similar in 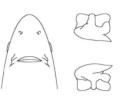 both jaws. These sharks are gray above, with conspicuous white spots present on their flanks, and lighter colored below; the fins are without white edges or other prominent markings. Tooth count: 26–29/22–26. Vertebral count: 97–106. Spiral valve count: 12–13.

HABITAT AND RANGE: Spiny Dogfish are perhaps one of the most abundant of all shark species. This rather gregarious shark forms large localized schools of hundreds if not thousands of individuals of uniform size and sex. They are found close inshore and offshore, on the continental shelf and upper slope, from the surface down to a depth of at least 1,236 m. These sharks appear to move in areas in which the water temperature ranges from 45 to 59 degrees F, often making longitudinal and depth migrations to follow this optimal temperature gradient.

Spiny Dogfish occur from the Gulf of Alaska, with isolated individuals occasionally found in the Bering Sea, southward to San Martin Island, in southern Baja California. They are extremely abundant in waters off British Columbia and Washington, but decline in abundance southward along the Oregon and California coasts. Elsewhere this antitropical shark is one of the most widely distributed species in both the Atlantic and Pacific Oceans.

NATURAL HISTORY: Viviparous, without a yolk sac placenta, with litters of up to 20, but most averaging between two and 12. Litter size and the size at birth are correlated with the female's size; larger females have more young and give birth to larger pups. Differences in litter number and size at birth vary regionally. The sex ratio at birth is 1:1. The gestation period is 18 to 24 months, the length of time depending on area. The breeding season is between September and January. Birth occurs in the midwater zone overlying depths of 165 to 350 m. The young tend to occupy a pelagic habitat, but shift to a more demersal lifestyle with maturity. Size at birth is 22 to 33 cm. Size at maturity varies regionally, but for eastern North Pacific Spiny Dogfish it is 70 to 80 cm for males and 80 to 100 cm for females. Maximum size for northeastern Pacific males is 107 cm and for females 130 cm. Elsewhere exceptional individuals may reach 160 cm, but most are smaller.

This is a very slow-growing, long-lived species. Age at maturity for eastern North Pacific Spiny Dogfish averages 14 years for males and 16 to 35 years for females, with an average age at maturity of 24 years. The age at maturity varies regionally between stocks. Maximum age is at least 30 to 40 years with some estimates reaching up to 100 years.

Spiny Dogfish are active, voracious predators that feed on invertebrates, including squids, octopuses, crabs, shrimps, gastropods, sea cucumbers, jellyfishes, and combjellies, as well as bony fish species such as herring, sardines, clupeids, smelt, hake, rockcod, flatfishes, sculpins, and almost any bony fish smaller than itself. Cartilaginous fishes are uncommon prey items with the exception of the White-Spotted Ratfish.

Spiny dogfish are opportunistic feeders, shifting their diet between species as prey abundance changes due to depth, locality, and time of year. Juvenile Spiny Dogfish are mostly pelagic, with a diet that consists primarily of small invertebrates, but that shifts mainly to fish as they grow and assume more of a bottom-dwelling existence. Adolescents and adults venture into the midwater and have been observed at the surface feeding on euphausiids, a kind of shrimp. During times of the year when herring are spawning these sharks consume not only herring, but the herring roe as well.

Spiny Dogfish are preyed upon by a number of larger shark species, some bony fishes including lancetfishes, and some rockfish species, pinnipeds, and Killer Whales. Shark predators on Spiny Dogfish include the Sixgill, Sevengill, Leopard, and Great White Sharks, which often have fin spines imbedded in their mouth or stomach wall.

Tag recapture of isolated individuals has revealed spectacular movements, suggesting that certain stocks, parts of stocks, or individuals are highly mobile. A large adult male tagged in British Columbia was recaptured off Santa Cruz six months later, a distance of 1,124 miles, and an individual tagged off Vancouver Island, British Columbia, was later recaptured off southern Baja California near the southern range limit for this species in the eastern Pacific, a distance of 2,441 miles. Several other recaptures of individuals tagged off Vancouver Island have been made at more moderate distances along the Oregon coast and as far south as San Francisco. Perhaps the most remarkable recaptures were two trans-Pacific specimens caught off Japan: one was recaptured off northern Honshu, Japan, seven years after release near Willapa Bay, Washington, and the other was recaptured off northern Hokkaido after only two years at liberty from a tagging off British Columbia. The minimum distances traveled for these two individuals were 4,902 miles and 4,156 miles, respectively.

HUMAN INTERACTIONS: The eastern North Pacific fishery for this species is centered in waters off British Columbia and Washington, where the fishery typically occurs during summer. Historically the commercial fishery in California, which peaked in the 1940s, occurred between Eureka and Fort Bragg. The California fishery for Spiny Dogfish usually took place during winter when Spiny Dogfish populations moved south. Presently there is no directed fishery for Spiny Dogfish in California.

Elsewhere this is one of the most commercially important shark species in cooler waters. In addition, this species is known to damage gear used for other target species as well as to consume other commercially important species. They are typically taken in bottom trawls or gill nets. They are utilized as food for human consumption as well as for liver oil, pet food, fish meal, fertilizer, and leather products.

These sharks are not dangerous in the sense of attacking people, but they may inflict severe injury to those who catch them due to their sharp teeth and mildly toxic dorsal fin spines. When captured this shark will curl itself and whip its tail about, inflicting injury from the long, sharp second dorsal fin spine.

NOMENCLATURE: *Squalus acanthias* (Linnaeus, 1758). The scientific name comes from the Latin *Squalus* meaning shark and the Greek *acanthias,* which refers to its dorsal fin spines. The common name Spiny Dogfish refers to the distinctive dorsal fin spines.

This shark has also been locally referred to as the Piked Dogfish, Spotted Spiny Dogfish, and Spur Dogfish.

In some accounts the eastern, North Pacific Spiny Dogfish *Squalus suckleyi,* is considered a distinct species from the North Atlantic, *S. acanthias.* The validity of this has been disputed as no significant morphological characteristics have been found to separate these species.

REFERENCES: Jones and Geen (1977a–c); Ketchen (1986); Saunders and McFarlane (1993).

Lanternsharks (Etmopteridae)

The lanternsharks comprise five genera and over 45 species, with numerous species awaiting formal description; one species is known to occur in California waters. These are mostly small, slender to cylindrical-shaped sharks, with a fifth gill opening that is noticeably smaller than the first four; their teeth are usually of similar length in both jaws, but if not, the lowers are usually longer; the first dorsal fin, which originates closer to the pectoral fins than the pelvic fins, is equal to or smaller than the second dorsal fin; each dorsal fin is preceded by a fin spine, the second usually longer than the first; the caudal peduncle lacks lateral keels and precaudal pits, the interdorsal space is usually longer than the dorsal fin bases, and the caudal fin is heterocercal shaped. Many sharks in this family have photophores on their lower body surface. This group contains some of the world's smallest living shark species, with most species less than 1 m in length and several species less than 50 cm in length at maturity. These are wide ranging, mostly deepwater, epipelagic, or benthic sharks found in continental and insular slopes or in the open ocean far from land.

PACIFIC BLACK DOGFISH   *Centroscyllium nigrum*

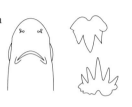

DESCRIPTION: A soft, flabby body with a moderately long abdomen; two similar-sized dorsal fins, each preceded by a fin spine, the second being 1.5 times as long as the first; and a short caudal peduncle. The teeth are similar in both jaws, with a large median cusp flanked by one or two smaller, smooth-edged cusplets on each side. Coloration is a uniform dark brown or black above, becoming darker below; dorsal, pectoral, and pelvic fins have white-tipped edges. Tooth count: 40–66/40–60. Vertebral count: 81–85. Spiral valve count: 4–6.

HABITAT AND RANGE: Very little is known about this shark other than that it occurs on or near the bottom at depths of 400 to 1,145 m. Specimens have been captured on soft mud or sand bottoms. Based on the presence of mesopelagic prey items in its stomach, this shark appears to migrate into the water column to feed.

The Pacific Black Dogfish has occasionally been caught in deep water off southern California, particularly around the Channel Islands. Elsewhere it is sporadically distributed throughout the central and eastern Pacific.

NATURAL HISTORY: Viviparous, without a yolk sac placenta, with a litter size of up to at least seven. Size at birth is 11 to 13 cm, with newborn specimens having an internal yolk sac for nourishment. Adult males are 35 to 39 cm and females about 43 cm; maximum size is approximately 50 cm.

Pacific Black Dogfish feed on deepwater shrimps, cephalopods, and small mesopelagic bony fishes.

HUMAN INTERACTIONS: These sharks are incidentally caught in sablefish traps set in deep water off southern California, but they are not utilized.

NOMENCLATURE: *Centroscyllium nigrum* (Garman, 1899). The genus name *Centroscyllium* comes from the Latin *centrum*, meaning prickly, and *scyllium,* meaning a kind of dogfish. The specific name *nigrum,* meaning black, is in reference to the color of this shark, hence the common name Pacific Black Dogfish. This shark is also locally referred to as the Combtooth Dogfish.

REFERENCES: Hubbs et al. (1979).

Sleeper Sharks (Somniosidae)

The sleeper sharks are among the world's largest sharks, with some species reaching over 5 m in length. This family is composed of four or five genera and 15 or more species, one of which occurs off California. These moderate to large-bodied sharks range from 1.5 to over 5 m long with five similar-sized gill openings; dissimilar teeth in the upper and lower jaw, the uppers being narrow with no cusplets or blades and the lowers being larger than the uppers with an erect to oblique cusp, with adjacent teeth being imbricated in the lowers but not the uppers; two small dorsal fins, with or without fin spines; the first dorsal fin may be closer to the pectoral fin bases than to the pelvic fin bases or vice versa; the caudal peduncle lacks lateral keels or precaudal pits, but the heterocercal-shaped caudal fin may have a small subkeel on it. These sharks lack photophores. This shark group is wide ranging and one of the few groups to contain true polar species. These sharks are found on continental shelves and slopes, around seamounts, occurring from very shallow polar to quite deep tropical waters.

PACIFIC SLEEPER SHARK   *Somniosus pacificus*

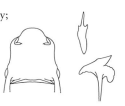

DESCRIPTION: A large, stout, flabby body; a relatively small transverse mouth; two small, spineless dorsal fins, the first originating midway on the trunk behind the free rear-tip of the pectoral fins; a broad, asymmetrical caudal fin with a subterminal notch and lateral keels on the fin base. The upper teeth have a single long, slender, erect, smooth-edged cusp; lower teeth have a short, low, strongly oblique, smooth-edged cusp. The color ranges from a uniform blackish brown to slate gray, without spotting or other distinctive markings. Tooth count: 33–48/48–60. Vertebral count: 39–43. Spiral valve count: 34–37.

HABITAT AND RANGE: The Pacific Sleeper Shark is a deep-living, bottom-dwelling shark known to occur at a depth of at least 2,000 m, at which depth they appear to be quite common off southern and Baja California. In the northern part of their range they may occur in extremely shallow water, at the surface, and even intertidally in Alaskan waters. In the southern portion of their range, including the California coast, these sharks are found in progressively deeper water. Juveniles have been captured in midwater trawls over very deep water.

The habitat of these sharks ranges from the Bering Sea southward to Baja California in the east and to Siberia and Japan in the west. Its northern distribution above the Arctic Circle is unclear as it may overlap that of the Greenland shark (*S. microcephalus*), a similar looking, but distinctive, large cold-water shark of the North Atlantic. Reports of Pacific Sleeper Sharks from the southern hemisphere appear to be a different *Somniosus* species.

NATURAL HISTORY: Viviparous, without a yolk sac placenta; litter size is unknown, but may be large based on records of females carrying over 300 ovarian eggs. Females mature at about 3.7 m and grow to at least 4.4 m, but are reputed to reach 7 m. Size of males at maturity is poorly known other than they mature by 4 m. Adult males have rarely been observed and may live in considerably deeper water than juveniles and adult females. The smallest free-swimming specimen measured 65 cm.

These sharks are generally considered bottom feeders, with a wide variety of prey items including halibut, rockfish, blackcod, cephalopods, crustaceans, pinnipeds, and cetaceans as carrion. They are also known to catch harbor seals, sea lions, and fast-swimming epipelagic fish, such as albacore and salmon. This rather sluggish looking shark may be able to catch fast-swimming prey by lying in wait and ambushing it. The morphology of the Pacific Sleeper Shark's jaw and buccal cavity suggests that it is a rather powerful suction feeder.

HUMAN INTERACTIONS: These sharks are occasionally taken as a by-catch in trawl nets. The flesh is relatively soft and of poor quality for human consumption.

Despite its great size and appetite for marine mammals, this deepwater shark has never been implicated in attacks on people.

NOMENCLATURE: *Somniosus pacificus* (Bigelow and Schroeder, 1944). *Somniosus* comes from the Latin meaning sleepy and *pacificus* refers to the Pacific Ocean where this shark is known to occur. The common name for this large, lethargic shark is derived from the Latin name.

This shark was first recorded in California waters in 1920, and at the time was considered synonymous with the Greenland shark, *S. microcephalus*. Not until 1944 was the Pacific population described as being distinct from the North Atlantic species.

REFERENCES: Ebert et al. (1987); Scofield (1920).

Kitefin Sharks (Dalatiidae)

The kitefin sharks are a small, diverse group with seven genera and nine species, one or two of which occur off California. These are slender to stout-bodied sharks ranging in size from less than 1 m to a maximum length of about 2 m; the fifth gill opening is not enlarged compared to the first four, except in the taillight shark (*Euprotomicroides zantedeschia*); the spiracles are large and set close behind the eyes. The teeth are dissimilar in the upper and lower jaws, the upper teeth being small, with a narrow cusp and no cusplets and the lower teeth broad, blade-like, and with erect to oblique cusps; the adjacent teeth are imbricate in the lowers, but not the uppers. Some species have photophores, whereas others have specialized glands on the shoulder or cloaca. The dorsal fins are small to moderate in size, the first usually smaller than the second, with the first originating closer to the pectoral fin bases than to the pelvic fin bases except in the Pygmy Shark and the Cookiecutter Shark. The second dorsal fin base is either over or behind the pelvic fin bases; the caudal peduncle may or may not have lateral keels; and the caudal fin is heterocercal shaped. These are wide ranging in warm-temperate and tropical seas, and are usually found in deep water on continental slopes and in the open ocean far from land.

1a   First dorsal fin inserting well anterior to pelvic fin bases; second dorsal fin base about four times the length of first . . . . Pygmy Shark (*Euprotomicrus bispinatus*)

1b   First dorsal fin inserting over pelvic fin bases; both dorsal fin bases are similar in size . . . . Cookiecutter Shark (*Isistius brasiliensis*)

PYGMY SHARK   *Euprotomicrus bispinatus*

DESCRIPTION: One of the smallest of all known shark species, this dogfish has a long conical snout; a small spineless first dorsal fin located closer to the pelvic fins than to the pectorals; a long, low spineless second dorsal fin; and a caudal peduncle with low lateral keels and a nearly

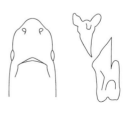

symmetrical caudal fin. The upper teeth have a single, small, narrow, erect, smooth-edged cusp. The lower teeth are larger, blade-like, and semierect. Coloration is black above, becoming lighter ventrally; the posterior fin margins are noticeably lighter.

Tooth count: 19–21/19–23. Vertebral count: 60–70. Spiral valve count: 12–13.

HABITAT AND RANGE: This is a warm-temperate to tropical, oceanic shark, usually found in areas in which the surface water temperature is 61 to 81 degrees F and the depth is 1,829 to 9,938 m. The Pygmy Shark is known to migrate to the surface at night, ascending from daytime depths below at least 500 m. All known captures of this species were made at the surface at night; no specimens have been captured in midwater trawls during daylight or at night. Most captures were of individuals that had been attracted to lights from boats. Circumstantial evidence suggests that this species may vertically migrate up to 1,500 m or more each way, which would put it out of range of most midwater trawls. Sand grains in the stomach of one individual found in water over 1,800 m deep indicate that it had been feeding on the bottom.

The Pygmy Shark is known from only a few specimens taken nearly 500 miles off southern California. A cosmopolitan species occurring mostly in the vast central water masses of the three major oceans, it tends to remain far from land, except for oceanic islands such as the Hawaiian Islands.

NATURAL HISTORY: Viviparous, without a yolk sac placenta, with litters of eight young. Males mature at 17 to 19 cm and grow to at least 22 cm. Females mature at 22 to 23 cm and grow to at least 27 cm. Size at birth is 6 to 10 cm. This is not only the smallest species off California, but one of the smallest known shark species.

The Pygmy Shark is an active, voracious feeder on midwater cephalopods and teleosts, including hatchfishes, lanternfishes, and lightfishes, and to a lesser extent on small crustaceans. Although they do consume fairly sizable cephalopods in relation to their size, Pygmy Sharks do not appear to take prey items as large as some of the squid eaten by the Cookiecutter Shark.

Pygmy Sharks possess luminescent organs on their ventral surfaces that may serve to camouflage them from predators when they are at the surface. These organs may also play an important role in feeding and social recognition.

HUMAN INTERACTIONS: None.

NOMENCLATURE: *Euprotomicrus bispinatus* (Quoy and Gaimard, 1824). The generic name comes from the Greek *Eu* and *protos,* meaning primitive, and *micrus,* meaning small. The specific name is from the Greek *bi,* meaning two, and *spinatus,* meaning spine, in

reference to its dorsal fin spines. This is one of the smallest known shark species, which accounts for its common name, Pygmy Shark.
REFERENCES: Hubbs et al. (1967).

COOKIECUTTER SHARK   *Isistius brasiliensis*

DESCRIPTION: A small, slender, cigar-shaped shark, with a bulbous snout; large oval eyes; a short, nearly trans-verse mouth; two close-set, similar-sized dorsal fins, the first originating just anterior to the pelvic fin bases and the second originating over the pelvic fin free rear-tips; and a large, nearly symmetrical caudal fin. The upper teeth are small, with a single 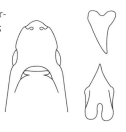 smooth-edged, nearly erect cusp. The lower teeth are triangular-shaped and are much larger and broader than the upper teeth. Coloration is a uniform brown above and slightly lighter below, with a prominent dark collar around the gills. The posterior fin tips have light edges, except for the caudal fin lobes, which are dark. Tooth count: 30–37/25–31.* Vertebral count: 81–89.* Spiral valve count: 8–10.*

HABITAT AND RANGE: An oceanic species generally found in tropi-cal and warm-temperate waters. They appear to be most com-mon between 20 degrees north and south latitudes, where the surface water temperatures range from 64 to 79 degrees F. This species, like the Pygmy Shark, is known to make tremendous daily vertical migrations of over 1,000 m. Most of the specimens that have been caught were taken at night, often near the surface. They are known to occur at a depth of at least 3,500 m.

Their presence in California waters is unconfirmed, although fresh wounds on marine mammals from as far north as Ano Nuevo and the Farallon Islands indicate that they likely do occur in this area, especially during warm-water years. The closest confirmed recorded specimen is from off Guadalupe Island, Mexico, a distance of little more than 240 miles from California. This circumglobal species is found in scattered locations throughout all of the major ocean basins.

NATURAL HISTORY: Viviparous, without a yolk sac placenta, with six to 12 young per litter. Males mature at about 36 cm, with a maximum length of 42 cm. Females mature at about 39 cm and reach a maximum length of 56 cm. Size at birth is unknown.

Cookiecutter Sharks are voracious, parasitic predators, feeding on midwater cephalopods, large fishes, and marine mammals, including tunas, dolphinfishes, billfishes, Megamouth Sharks, Deepsea Stingrays, Northern Elephant Seals, fur seals, and cetaceans. They have also been known to take plug bites out of undersea cables and the rubber sonar domes on nuclear submarines.

Cookiecutter Sharks have specialized suctorial lips that enable them to attach to their victim prior to removing its flesh. It is believed that a band of pigmentation located beneath the jaw and bounded by ventrally directed bioluminescent pores acts as a lure to attract large, active swimming pelagic predators. As the predator closes in on what it believes to be a potential prey item, the Cookiecutter Shark turns the tables by attacking and parasitizing it. The Cookiecutter stabs the predator using its sharp lower teeth and suctorial lips at the moment of contact to form a suctionlike grip on its victim. Its body then contracts simultaneously to create a twisting half-turn movement that leaves a craterlike wound. The predator in effect becomes the prey. Because of the victim's tremendous size its resulting wound is probably little more than an irritant. Smaller prey items such as cephalopods are consumed entirely by these sharks. Cookiecutter Sharks may travel in large schools when foraging for food.

HUMAN INTERACTIONS: Although sketchy, there are records of people being attacked by small sharks in the open sea. An underwater photographer reported being attacked by a swarm of small sharks, less than 60 cm long, while diving in the open sea. There are records of shipwreck survivors who reported being attacked at night by very small sharks that inflicted deep, clean bites of an inch or so. Generally, however, Cookiecutter Sharks are not considered dangerous to humans, as they tend to be found far offshore and are among the smallest known shark species.

Cookiecutter Sharks are of no commercial value and may be caught on occasion as a by-catch in other fisheries.

NOMENCLATURE: *Isistius brasiliensis* (Quoy & Gaimard, 1824). The generic name *Isistius* is of uncertain origin. The specific name *brasiliensis* comes from the country of Brazil, off whose waters the holotype was caught. The common name comes from the spherical plugs it takes out of its prey. These plugs are reminiscent of those made by a household cookiecutter.

REFERENCES: LeBouf et al. (1987); Widder (1998).

ANGEL SHARKS (SQUATINIFORMES)

The angel sharks, which comprise a small, undiverse order, are all included in a single family and genus. There are 15 species recognized worldwide with several others awaiting formal description. A single species occurs along the California coast. This group has adapted to bottom living to such a degree that the bodies of its members appear raylike in form with a mottled dorsal surface and large pectoral fins that extend over the gill slits. Unlike rays, however, in which the gill slits lie completely on the ventral surface and the pectoral fins are fused to the head, the gill slits of angel sharks lie at least partly lateral to the head and the pectoral fins are not fused to the sides of the head. Angel sharks are so distinctive in appearance that there is little likelihood of confusing them with any other sharks. Their maximum size is 2.4 m, but most are less than 1.5 m long.

Angel Sharks (Squatinidae)

These are cold-temperate to tropical sharks often seen lying partially buried on the bottom in mud or sand. They are usually found on continental shelves and slopes from the intertidal zone to a depth of 1,390 m. Members of this order may have litters of up to 25 young. Angel sharks are voracious feeders, consuming a variety of fishes and squid.

PACIFIC ANGEL SHARK   *Squatina californica*

DESCRIPTION: A distinctive, flattened, raylike shark, with large pectoral fins clearly separated from the head; a terminal mouth that is nearly three-fourths the head length; and gills that are situated laterally, close together, and anterior to the pectoral fins. The teeth are similar in both jaws with a single, large, sharply pointed cusp. Their body coloration ranges from gray to brown or reddish above with scattered dark spots and white below. Tooth count: 14–18/14–18. Vertebral count: 140–141. Spiral valve count: 7–9.

HABITAT AND RANGE: The Pacific Angel Shark is a cold to warm-temperate raylike shark found on the continental shelf usually at depths of 3 to 100 m, but on rare occasion it occurs down to at least at 183 m. They are often found in shallow bays, in estuaries, at the head of submarine canyons, and around rocky reefs and kelp forests. Although primarily a bottom-dwelling species, these sharks have been reported to swim 15 to 91 m off the bottom. Pacific Angel Sharks do not appear to make long-distance migrations and in fact the deep ocean basins separating the northern and southern Channel Islands as well as the mainland act as barriers that limit their movements.

The eastern North Pacific angel shark population is composed of several discrete subpopulations, some of which are so distinct from each other that they may in fact represent an entirely different species. Within the Southern California Bight at least three discrete subpopulations have been identified—one each in the southern and northern Channel Islands and a third along the mainland. North of Point Conception to southeastern Alaska they occur in several discrete isolated populations along the coast. To the south in Mexican waters there appears to be at least two additional subpopulations, one along the Pacific coast of Baja and a second in the Gulf of California. The populations between the Pacific coast and the Gulf of California, although considered synonymous, are so distinct that they may in fact represent different species.

They are endemic to the eastern Pacific from southeastern Alaska to the Gulf of California and from off Ecuador to southern Chile, although the South American variety may be a different species.

NATURAL HISTORY: Viviparous, without a yolk sac placenta, with litters ranging from one to 13 young, averaging six. Unlike some sharks the number of young does not increase in larger females. Birth takes place between March and June following a 10-month gestation period. At birth they are about 25 cm. Males mature at about 1 m and females at 1 to 1.2 m. Maximum length for males is 118 cm and for females is 152 cm. The population in the Gulf of California matures at a much smaller size—78 cm for males and 85 cm for females—than the population along the Pacific coast, strongly suggesting that it may in fact be a different species.

Angel Sharks of both sexes mature in about 10 years and may live up to 35 years. The average growth rate is 15 to 19 cm per year for young Angel Sharks, but slows to around 2 cm or less per year for adults.

Angel Sharks feed primarily on bony fishes, including kelp bass, croaker, white sea bass, blacksmith, flatfish, queenfish, mackerel, and Pacific sardines, from spring through early winter, but shift to spawning squid, which predominate during winter and early spring. Newborn Angel Sharks have an internal yolk sac that provides nourishment during the first month or so of life.

Once this yolk sac has been used up, however, these newborns must feed on their own.

Angel Sharks are ambush predators, lying in wait until an unsuspecting prey item swims within 4 to 15 cm of them before quickly striking. With a swift upward head lunge combined with a thrust of their protrusible jaws, these sharks will then snap up the swiftest swimming of the prey. Vision appears to play an important role in prey capture. Prey items more than 15 cm away have a higher probability of surviving an ambush as the Angel Shark is less accurate in its attack from this distance. After attacking and consuming a prey item, these sharks quickly position themselves to await their next victim. They resettle by using their pectoral fins to create a depression so that they can approximate the level of the bottom and by leaving a thin covering of sand on their backs to disguise their outline.

Angel Sharks tend to lie on soft, flat bottoms near areas of vertical relief such as rock–sand interfaces or rocky outcrops. Because these rocky reefs serve as a prime refuge area for a wide variety of fishes, such a location enhances the chance of encountering potential prey. As these fishes venture away from the reef to forage or migrate, they are ambushed. Angel sharks will remain in virtually the same location for extended periods of time if they are successful in ambushing prey. These ambush sites are actively selected based on the sharks' hunting success.

Angel sharks are preyed upon by larger sharks, including the Great White Shark and Sevengill Shark.

HUMAN INTERACTIONS: A commercial fishery for angel sharks emerged in the late 1970s off Santa Barbara and continued until the mid-1990s when drift gill nets were banned by the state, effectively ending this fishery. The fishery peaked in the mid-1980s with landings exceeding 1 million pounds, making the angel shark the number one species of shark caught in California waters. The meat of angel sharks is of high quality for human consumption and is marketed fresh or frozen. These sharks are also taken in some numbers by recreational anglers, particularly off southern California.

Angel sharks are not considered dangerous and are easily approached by divers, but caution should be exercised as they can inflict a painful bite when provoked.

NOMENCLATURE: *Squatina californica* (Ayres, 1859). The genus name *Squatina* is from the Latin meaning skatelike and the species name *californica* is in reference to where it was first described. The common name, Pacific Angel Shark, refers to its occurrence in the eastern North Pacific. The Pacific Angel Shark is frequently referred to simply as the Angel Shark, California Angel Shark, or Monkfish.

The angel shark was originally described by Ayres (1859), although Jordan and Evermann (1896) considered it to be synonymous with the European *S. squatina*. The systematics of eastern Pacific angel sharks is poorly understood. An eastern South Pacific form may be identical to this species or may be a distinct species, *S. armata*. Closer to our area, an angel shark from the Gulf of California referred to as *S. californica* may in fact be an undescribed species. Supporting evidence for two distinct species comes from the fact that Gulf of California angel sharks mature at a much smaller size than those found on the Pacific coast of Baja. A careful taxonomic review of the *Squatina* species from the eastern Pacific is required to clarify the validity of these nominal species.

REFERENCES: Fouts and Nelson (1999); Natanson and Cailliet (1986, 1990); Strong (1989); Villavicencio-Garayzar (1996a).

HORN SHARKS (HETERODONTIFORMES)

The horn or bullhead sharks are a minor shark group with one family and a single genus of eight similar looking species. These are the only living sharks with a spine preceding each dorsal fin and an anal fin. Horn sharks are small- to medium-sized with a stout body, tapering posteriorly from a broad, blunt head; a short piglike snout; a broad crest over each eye; and a short transverse mouth with small cuspidate teeth in front and enlarged flattened crushing teeth in the rear. Some species are reported to reach 1.65 m, but most are less than 1 m in length.

Horn Sharks (Heterodontidae)

Three species of horn sharks are known from the eastern Pacific, but only one occurs in California waters. Horn sharks are found on the continental shelves and upper slopes of warm-temperate and tropical seas, usually in areas in which the water temperature is above 70 degrees F. They are benthic-living, shallow-water sharks typically found from the intertidal zone out to a depth of 100 m, although one species, found in the western Indian Ocean, is unusual in that it typically is found at a depth of 100 to 275 m. Development is oviparous, with the members of this group laying distinctive hand grenade–shaped egg cases with spiral flanges. Their diet consists primarily of benthic invertebrates, including polychaete worms, molluscs, and echinoderms.

HORN SHARK FOLLOWS ➤

HORN SHARK

Heterodontus francisci

DESCRIPTION: The Horn Shark is the only shark that combines two spined dorsal fins and an anal fin. The teeth are strongly differentiated with smaller, single-cusped, anterior teeth and larger molarlike posterior teeth in both jaws.

The body coloration ranges from gray to brown with light to dark shades of each color above. The ventral surface is yellowish and the dorsal surface has a scattering of small black spots, although large specimens may lack spotting. Tooth count: 19–26/18–26. Vertebral count: 112. Spiral valve count: 8.

HABITAT AND RANGE: The Horn Shark is a warm-temperate to subtropical benthic shark found on the continental shelf from the intertidal zone out to a depth of 152 m, although it is most common at a depth of 2 to 11 m. During winter, Horn Sharks migrate into deeper water usually below 30 m. Their preferred habitat changes as they develop and mature. Adults prefer rocky reefs with caves and crevices, or areas of thick algae cover, whereas juveniles inhabit sandy bottoms with little vertical relief. Adult Horn Sharks that favor an algal habitat have noticeably longer fin spines than those preferring a reef habitat. Those residing on rocky reefs tend to wear down their fin spines going into and out of caves and crevices.

Adult Horn Sharks are fairly inactive during the day but become quite active at night. They are site specific, returning to the same resting location at dawn and remaining there until the following evening. Juveniles tend to be more active during the day. Horn Sharks have a narrow home range, usually no more than

1,000 m^2, with some having been observed to remain in the same localized area for over 11 years.

Water temperature is an important factor controlling the relative abundance of Horn Sharks. Studies comparing the population densities of Horn and Swell Sharks at Catalina Island have shown that the Horn Shark population had increased over a 20-year period whereas the Swell Shark population had decreased during the same time. This change in population density has been attributed to an increased warming of the surface waters in southern California. Both the Horn and Swell Sharks occupy a relatively similar habitat in areas they co-inhabit. Horn Sharks prefer water temperatures above 70 degrees F whereas Swell Sharks can tolerate slightly cooler water temperatures. Swell Sharks are more abundant and Horn Sharks are less abundant in the cooler waters of the northern Channel Islands. It is of interest that the holotype of the Horn Shark was taken in Monterey Bay in the 1850s at a time of unusually warm water. Other than the holotype, there are no confirmed records of the Horn Shark this far north.

Horn Sharks are endemic to the eastern Pacific as they occur only from southern California to the Gulf of California, and possibly to Ecuador and Peru, although this is unconfirmed.

NATURAL HISTORY: Oviparous, with mating taking place in December and January. Females lay two egg cases every 11 to 14 days usually between February and April. Over the course of a breeding season a single female may deposit up to 24 egg cases. Development lasts six to nine months depending on the water temperature. These spiral-flanged egg cases are light brown when first laid, but darken after a few days. Differences in egg case size between the populations of the Channel Island and the mainland suggest that these two populations have been isolated from each other for some time. Horn Shark egg cases found around the Channel Islands tend to be slightly longer than those found along the mainland. Egg cases are usually laid in relatively shallow water (between 2 and 13 m deep). These are the only sharks that appear to exhibit any form of parental care, as the females will carry the egg cases in their mouth and wedge them under rocks or in crevices. The young at birth are about 15 to 17 cm with their size at birth corresponding to the size of the egg case. Males mature at 56 to 61 cm and reach a maximum length of 83 cm. Females over 58 cm long are mature, with a maximum length of at least 96 cm, but possibly to 120 cm, although this is unconfirmed.

Horn Sharks exhibit a high degree of segregation corresponding to their life history. Juveniles tend to occupy a relatively shallow, sandy habitat whereas adults prefer either rocky reefs or dense algal bottoms. Adolescent Horn Sharks, between 35 and 48 cm long, tend to remain in deeper water, usually between 40 and 150 m. As they reach adulthood Horn Sharks migrate back into relatively shallow water. This separation of habitat by size and stage of maturity reduces competition for food and habitat between younger and older sharks. This is probably good as some Horn Sharks are known to feed on the egg cases of their own kind.

Very little is known regarding the age and growth of Horn Sharks. Their growth rates are generally slow and extremely variable, and do not correspond to their size. Based on tagging studies, growth rates were found to range from 0 to 6 cm per year. There is one unconfirmed account of a Horn Shark living up to 25 years.

Adult Horn Sharks feed on a wide variety of benthic invertebrates including marine snails, crabs, shrimps, squids, sea urchins, sea stars, and small bony fishes. Larger Horn Sharks with a preference for sea urchins, particularly the Short-Spined Purple Urchins, develop a purple tint to their teeth and fin spines. Horn Sharks in southern California will switch their prey preference between winter, when squid are especially abundant, and summer, when fishes, particularly blacksmiths, are far more abundant. Juveniles tend to prefer polychaete worms, small clams, and sea anemones but will opportunistically feed on squid and small bony fishes when available.

Adult Horn Sharks usually shelter themselves at the base of reefs, in caves, in crevices, or in algae beds until shortly after dark when they move up the reef slopes in search of food. They will forage along reef tops until dawn when they return to the same daytime shelter. Their daytime sheltering sites are usually located peripheral to their nighttime foraging range. Unlike Angel Sharks, which will swim 15 to 91 m off the bottom at times, Horn Sharks orient themselves to the bottom, rarely moving more than 2 m from it.

Juvenile Horn Sharks, which occupy a sandy habitat similar to that of Bat Rays, have developed an interesting relationship with their flattened relative. Feeding Bat Rays will raise and lower their body in a "push-up"-like manner by using their pectoral and pelvic fins for support. This rapid, repetitive motion creates depressions in the sediment, exposing polychaete worms and small clams, both of which the Bat Ray consumes. These depressions left by Bat Rays are utilized by juvenile Horn Sharks as shelter on an otherwise featureless bottom and as a source for food by exposing potential prey items. Foraging Horn Sharks feed in these depressions by moving their head up and down in a pecking motion searching for prey items on the soft, sandy bottom.

It has been suggested that newborn Horn Sharks do not feed for about one month after they are born. Instead they survive by receiving nutrition from an internal yolk sac; when this has been used up they begin to actively forage for prey. However, newborns in captivity will begin feeding almost immediately after birth. It may be that juveniles, although receiving nourishment from an internal yolk sac, can also begin feeding orally if food is present.

Potential predators on Horn Sharks appear to include larger sharks and bony fishes, as well as pinnipeds. However, the tough, leatherlike skin and prominent dorsal fin spines of Horn Sharks make them somewhat undesirable prey items. Juveniles seem to be most vulnerable to predation as they tend to occupy an exposed sandy bottom habitat. Angel sharks would seem to be likely predators on juveniles because they occur in similar habitats and both are nocturnal in their behavior. However, the combination of camouflage coloration and fin spines offers a defense mechanism from these would-be predators. Additionally, small Horn Sharks tend to occur in slightly shallower water than angel sharks. Horn Shark egg cases are attacked by predatory boring marine snails that actively drill holes in the cases to extract the protein-rich yolk.

HUMAN INTERACTIONS: Horn Sharks are of no commercial value. They are taken as a by-catch in traps and trawls and occasionally by recreational anglers. Divers can easily approach these sharks and they are generally harmless unless harassed, in which case they may bite their antagonist. Care should be exercised with their spines, which can inflict a painful sting.

Horn Sharks are quite hardy and are commonly maintained in public aquariums, although given their nocturnal behavior pattern they do not display very well, often hiding from the public during the daylight hours. Individual specimens have been maintained in public aquariums for over 12 years.

NOMENCLATURE: *Heterodontus francisci* (Girard, 1854). The generic name *Heterodontus* comes from the Greek *heteros,* meaning different, and *odontus,* meaning tooth. The species name *francisci* is named for San Francisco, even though these sharks do not occur that far north. The common name refers to its dorsal fin spines. Occasionally, it has been referred to as the California Horn Shark.

The species was originally described as *Cestracion francisci* by Girard (1854) from a specimen collected in Monterey Bay at the northern extreme of its range. Gill (1862) later placed it in the genus *Gyropleurodus*. However, both *Gyropleurodus* (Gill, 1862) and *Cestracion* (Cuvier, 1817) are junior synonyms for *Heterodontus,* as originally described by Blainville (1816).

REFERENCES: Dempster and Herald (1961); Nelson and Johnson (1970); Segura-Zarzosa et al. (1997); Strong (1989).

CARPET SHARKS (ORECTOLOBIFORMES)

The carpet sharks are composed of seven families, 14 genera, and 31 to 34 species, one of which occasionally occurs in California waters. These are small to giant sharks with one species, the Whale Shark, being the largest living fish on earth. Carpet sharks can be distinguished by a short to truncated snout with a terminal or subterminal mouth connected to the nostrils by prominent nasoral grooves; distinct barbels on the inner nostril edges; two spineless dorsal fins; and an anal fin. Many members of this shark group have rather striking color patterns. These are tropical to warm-temperate sharks found from the intertidal zone out to a depth of about 200 m. They are most abundant in the western Indo-Pacific, but a few species are fairly wide ranging. Development may be oviparous or by aplacental viviparity.

Whale Sharks (Rhincodontidae)

Whale sharks, a monotypic family with a single species, are most notable for being the largest living fish, reaching up to 18 m in length. This is an unmistakable shark with its checkerboard color pattern; a broad, flat head, with a wide terminal mouth; long straight gill openings; prominent longitudinal ridges on its back; and a huge crescent-shaped tail. It is a global species occurring most commonly in warm-temperate to tropical seas.

WHALE SHARK FOLLOWS ➤

WHALE SHARK

Rhincodon typus

DESCRIPTION: The largest fish in the world, this enormous shark with a slightly flattened head has a terminal mouth; long, straight gill openings; prominent ridges along its flanks; and a prominent pattern of white  spots between vertical and horizontal stripes reminiscent of a checkerboard pattern. This unique pattern is offset by a background color that ranges from a dark gray to bluish or brown above, becoming white below. Tooth count: 300+/300+. Vertebral count: 153.* Spiral valve count: 68–73.*

HABITAT AND RANGE: The Whale Shark is an oceanic species usually found at or near the surface from close inshore, including bays and lagoons, to far out at sea. It is usually found in nearshore waters where the temperature is 64 to 86 degrees F, with a preference for mixing layers of cool upwelled oceanic water. Whale Sharks will dive to depths in excess of 240 m where the water temperature reaches 50 degrees F or cooler. They are highly mobile and may range for thousands of miles, often migrating seasonally to specific feeding locations. They will migrate in groups of similar size and sex, with movement patterns seeming to correlate with oceanographic features such as sea mounts and boundary currents where their preferred food items are abundant. When migrating they are often seen traveling in groups numbering from only a few up to 100 or more individuals. Migrating Whale Sharks average around 15 miles per day with occasional travel rates of up to 57 miles per day. Adults will migrate over very long distances, over 7,000 miles, whereas smaller juveniles exhibit a more localized movement pattern.

Whale Sharks are found throughout the eastern Pacific from off Santa Cruz in central California, where they are rare, to northern Chile and the Galapagos Islands. Small numbers of these sharks have been observed off San Diego, in the Santa Barbara Channel, and in Monterey Bay, with one record of this shark from as far north as Patrick's Point in Humboldt County, some 400 miles north of Monterey Bay. Whale Sharks are very common in the Gulf of California. A circumglobal species, Whale Sharks are found in tropical and warm-temperate seas.

NATURAL HISTORY: Viviparous, without a yolk sac placenta, with litters of at least 300, making it the most fecund species of cartilaginous fish. Interestingly, the embryos are retained within a leathery egg case inside the uterus, emerging from this egg case just prior to birth. This form of aplacental viviparity development is unique among chondrichthyans. Whale Sharks do not have a defined birthing season as newborns are found year-round in tropical areas. The size at birth is 55 to 64 cm. Males mature at 8.5 to 9 m and females at greater than 10 m. The largest reliably measured specimens were a male 12.2 m and a female 18 m in length.

Whale Sharks grow quite rapidly the first several years of life, averaging 20 to 30 cm annually. A single newborn held in captivity more than doubled its length in 90 days growing from 60 to 126 cm. Males mature in about 20 years and live at least 31 years. Immature females measuring 8 to 9 m are estimated to be 18 to 27 years old. It has been hypothesized, without supporting evidence, that these sharks may live up to 100 years.

Whale Sharks have a varied diet that includes plankton, copepods, jellyfish, invertebrate larvae, crustaceans such as pelagic crabs, krill, squid, and small- to medium-sized fishes including anchovies, sardines, mackerels, small tunas, and albacore. When targeting dense patches of their primary prey they at times forage in groups, often with other fishes and seabirds, numbering in the thousands. Whale Sharks will also aggregate when other fishes or corals are spawning to feed on the gametes as they are being released. These sharks will often appear in the same locations seasonally, especially when plankton patches, spawning activity, or other intense ecological processes involving preferred prey species are ongoing. In the Gulf of California, Whale Sharks appear seasonally to feed on plankton blooms near the Bay of La Paz. Younger Whale Sharks tend to move closer to shore to feed, whereas large, possibly adult animals remain further offshore. Newborn Whale Sharks have an internal yolk sac from which they receive nourishment in the first few weeks of life, after which they begin to feed orally on plankton.

Whale Sharks are suction filter feeders, drawing in water at a higher velocity than the more passive basking shark. They have dense gill rakers that act as sieves to screen out plankton and other prey items. They feed by rising vertically to the surface through dense patches of prey with their head raised nearly 1 m out of the water and then sinking slowly back down to water level with their mouth agape, creating a vacuum to inhale their prey. Then by rising back up, water flows out through their gills, which act as sieves to trap prey. They also feed by swimming horizontally through dense prey patches, mouth agape, and essentially vacuuming a swath through the concentrated prey. Whale Sharks are very dependent on dense concentrations of prey. In captivity, they consume, on average, 8 to 11 percent of their body weight in food per week.

Adult Whale Sharks have very few known predators. Only the Killer Whale is known to attack the adults. However, small juveniles (less than 1 m in length) are more vulnerable; they are known to be fed upon by the Blue Shark and the Blue Marlin, and it would not be unexpected for them to fall prey to other large predators. Because very few Whale Sharks less than 4 m in length have ever been reported, it may be that their rapid growth rate is an important mechanism of defense from possible predation by larger sharks.

HUMAN INTERACTIONS: Whale Sharks are extremely rare in California waters, being more common off the southern tip of Baja and in the Gulf of California. Boat charters operating off La Paz, Baja, take divers out to areas in which Whale Sharks are known to congregate, for a chance to dive with these magnificent creatures. They are fairly docile and will often allow divers to hitch a ride by hanging on to their dorsal or caudal fins. Whale Sharks have been kept in captivity for over five years.

NOMENCLATURE: *Rhincodon typus* (Smith, 1828). The name *Rhincodon* comes from the Greek *rhine,* meaning file, and *odon* (*odon*), meaning tooth, in reference to its many small teeth. The species name comes from the Greek *typos,* meaning shape or form, in reference to the tooth bands that resemble a file. The common name, Whale Shark, is in reference to the huge size of this shark.

The generic name *Rhincodon* is actually a misspelling of the original name *Rhiniodon,* which is often seen in the literature. Confusion over this spelling error was finally resolved by the International Commission on Zoological Nomenclature in 1984 when it ruled that in the interest of taxonomic stability *Rhincodon* was the correct spelling of the generic name.

REFERENCES: Chang et al. (1997); Clark and Nelson (1997); Colman (1997); Eckert and Stewart (2001); Joung et al. (1996); Winter (2001).

MACKEREL SHARKS (LAMNIFORMES)

The mackerel sharks are a small but diverse group containing seven very distinct families, 10 genera, and about 16 species of large oceanic and coastal sharks; six families, eight genera, and 11 species are found in California waters. Superficially, each of these shark families appears to be unique and perhaps unrelated, as this group contains such bizarre species as the goblin shark, with its acutely pointed snout; the giant filter-feeding megamouth shark; and the thresher sharks, with their elongated caudal fins nearly as long as their entire body trunk. The members of this order, however, are all united by a number of features, including a pointed snout, the lack of dorsal fin spines, and a similar style of dentition. This group on the whole has a wide circumglobal distribution from subpolar waters to the tropics. These sharks occupy a wide variety of habitats from nearshore coastal waters to an oceanic or deep-sea environment. They share a unique form of aplacental viviparity in which embryonic sharks feed on ovulating eggs (oophagy) and in some cases their fellow embryos within the uterus (intrauterine cannibalism).

1a Snout extremely elongated, flat, and bladelike. Precaudal pits and ventral caudal lobe absent ....
...............................goblin sharks (Mitsukurinidae)

1b Snout short to moderate, but not greatly elongated or bladelike, and broadly rounded. Precaudal pits and ventral caudal lobe present 2

2a Mouth terminal, lower jaw extending to snout tip ....
.................megamouth sharks (Megachasmidae)

2b Mouth subterminal, lower jaw not extending to snout tip
....3

3a Caudal fin about as long as rest of shark..................
..................................thresher sharks (Alopiidae)

3b Caudal fin much shorter than rest of shark..............4

4a Upper precaudal pit present; lower pit and lateral keels absent. Caudal fin heterocercal, not crescent shaped..
....................sand tiger sharks (Odontaspididae)

4b Upper and lower precaudal pits and strong lateral keel present. Caudal fin crescent shaped.................5

5a   Teeth large, triangular, and bladelike. Gill slits large, but not extending onto dorsal surface of head or nearly across throat, and without internal gill rakers . . . .
. . . . mackerel sharks (Lamnidae)

5b   Teeth minute, hook shaped, but not bladelike. Gill slits extremely large, dorsally extending onto surface of head and the first nearly extending across the throat, and with internal gill rakers . . . . basking sharks (Cetorhinidae)

Sand Tiger Sharks (Odontaspididae)

The sand tiger sharks are a small group of lamnoids composed of two genera and three or four species, one of which occurs in California waters. These sharks are characterized by a moderately stout, cylindrical body shape with a short head and a moderately long, pointed snout. The eyes, depending on the species, are small to moderately large; the mouth is large and located subterminal on the head; the teeth have a single large, long, narrow cusp flanked on either side by one or more cusplets. The teeth are an especially prominent feature of these sharks as they tend to protrude from the mouth in a viperlike fashion. Sand Tiger Sharks are a group of large sharks that range in size from 2.2 to 3.6 m in length. Depending on the species, they inhabit areas ranging from the coast, including bays and harbors, to very deep water.

SAND TIGER SHARK   *Odontaspis ferox*

DESCRIPTION: A large, stout-bodied shark with a conical to slightly flattened snout; a long mouth extending beyond the eyes; a first dorsal fin that is larger than the second dorsal and anal fins, originating over the pectoral fin free

rear-tips. The prominent teeth of this shark are long and narrow with a central cusp, flanked by two or three smaller cusplets on each side. The upper anterior and lateral teeth are separated by three to five smaller intermediate teeth. The body color is gray to brownish-gray or olive above and lighter below. Some specimens have dark reddish spots scattered over the body. Tooth count: 46–56/36–48. Vertebral count: 177–183.* Spiral valve count: 32.

HABITAT AND RANGE: The few records from California waters have been taken at a depth of 13 to 200 m, although this species is known to occur at depths of at least 420 m elsewhere. These sharks are typically found on or close to the bottom in deep water along the continental and insular shelves and upper slopes. There are several records of this species occurring pelagically in open ocean waters.

The Sand Tiger Shark is found from southern California southward along the Baja coast to the Gulf of California. There have been at least five confirmed recordings off southern California. Elsewhere it is known from a few scattered records in temperate and tropical seas in the North Atlantic, Indian, and Pacific Oceans and from the Mediterranean Sea, where it appears to be most abundant.

NATURAL HISTORY: Virtually nothing is known regarding the reproductive biology of this shark. It is assumed to be oophagous like other members of this group. Males mature at 275 cm and females at about 360 cm. The maximum reported size is 367 cm. The estimated size at birth is about 100 cm.

Sand Tiger Sharks feed mainly on small bony fishes, including rockfish, squids, and crustaceans.

HUMAN INTERACTIONS: The Sand Tiger Shark is too rare to be of any importance in California. Elsewhere this uncommon shark is taken in the Mediterranean Sea, the Gulf of California, and off Japan as a by-catch in bottom gill nets and long-lines and bottom trawls. It is used for human consumption, although its meat is considered of inferior quality. Its liver, which is very large and oily, has a high squalene content and probably serves this shark in a hydrostatic capacity.

The Sand Tiger Shark has never been implicated in an attack on humans, although care should be taken by those who may occasionally come into contact with this shark. Its large size and prominent teeth suggest that it may inflict serious injury if mishandled. The rarity of this shark in California waters combined with its deepwater habitat make it unlikely that most people will come into contact with it.

NOMENCLATURE: *Odontaspis ferox* (Risso, 1810). The generic name derives from the Greek *Odontos,* meaning tooth, and *aspis,* meaning viper, in reference to the fearsome appearance of these sharks. The Latin name *ferox* means fierce. The common name, Sand Tiger, is in reference to this shark's rather fearsome looking appearance. Other local common names include Sand Shark, Smalltooth Sand Shark, and Ragged-tooth Shark.

The Sand Tiger Shark was first identified from California waters as *Carcharias ferox*. The use of the genus *Carcharias* was based on a ruling by the International Commission on Zoological Nomenclature in 1912. This ruling was later overturned by the Commission in favor of *Odontaspis,* which was first proposed in 1838.

REFERENCES: Daugherty (1964); Seigel and Compagno (1986); Villavicencio-Garayzar (1996b).

Goblin Sharks (Mitsukurinidae)

The goblin sharks are an unmistakable, monotypic group, composed of a single, distinctive species. This is a bizarre looking shark with an elongated, flat, bladelike snout; a slender, soft flabby body; small eyes; and long, narrow, single-cusped teeth. The body coloration of freshly caught specimens is a pinkish-white with bluish fins, but the color fades shortly after death to a uniform brown. This is a poorly known deepwater species with a scattered distribution in both temperate and tropical seas. Goblin sharks are a fairly large species growing to nearly 4 m in length. Virtually nothing is known about their life history. The jaws are highly protrusible and most likely play an important role in the ability of this shark to catch its prey.

GOBLIN SHARK FOLLOWS ➤

GOBLIN SHARK

Mitsukurinia owstoni

DESCRIPTION: The Goblin Shark is un-mistakable, with its soft, flabby body; elongated, flat, bladelike snout; small eyes; and long narrow single-cusped teeth. The jaws of these sharks are highly protrusible. The body color of freshly caught specimens is a spectacu-lar pinkish-white with bluish fins. Pre-served specimens are uniformly brown

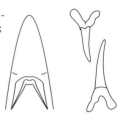

or gray. Tooth count: 38/38. Vertebral count: 122–124.* Spiral valve count: 19.*

HABITAT AND RANGE: The Goblin Shark is a poorly known deep-water shark found on the outer continental shelf and upper slopes down to a depth of at least 1,300 m. It is known from the eastern Pacific from a single specimen caught off San Clemente Island. Elsewhere this species is known from less than 50 specimens with records scattered throughout the three major oceans. The major-ity of specimens have been reported from Japanese waters.

NATURAL HISTORY: This species is assumed to exhibit aplacental viviparity with oophagous embryos as in other lamnoids, but this has yet to be confirmed. Males are known to be mature at 264 cm and to reach at least 385 cm. The largest recorded female was an immature specimen that measured 373 cm. The smallest known free-swimming specimen measured 107 cm.

Goblin Sharks feed on small pelagic bony fishes, pelagic crabs, and cephalopods. Based on their main prey species the Goblin Shark appears to forage away from the bottom for its food and may in fact occupy more of a midwater habitat than is generally assumed.

The Goblin Shark, with its flabby body and small fins, is a fairly inactive swimmer and probably relies on its rather special-ized features, the snout and jaws, for capturing prey. The long, bladelike snout covered with sensory cells, combined with its protrusible jaws and grasping teeth, suggests that this shark hunts its prey by detecting electric fields. Its large oily liver makes it nearly neutrally buoyant, thus allowing it to approach an unsuspecting prey item with relatively little movement. Once a prey item has been located, the Goblin Shark probably drifts close, so that its highly protrusible jaws can be rapidly projected to capture it. The combination of these specialized features and the dark, harsh environment in which it lives suggests that the Goblin Shark is an ambush predator.

The Blue Shark is a known predator of this species.

HUMAN INTERACTIONS: The Goblin Shark is of no importance other than it is occasionally taken as a by-catch.

A rare, bizarre-looking deepwater shark, it is relatively harmless to people. Goblin Shark teeth have been found imbedded in transoceanic undersea cables.

NOMENCLATURE: *Mitsukurinia owstoni* (Jordan, 1898). The family and generic names are in honor of Kakichi Mitsukuri. The specific name is in recognition of Alan Owston, who collected the type specimen. The common name comes from the bizarre appearance of this shark.

REFERENCES: Ugoretz and Seigel (1999).

Megamouth Sharks (Megachasmidae)

The megamouth sharks are a monotypic family of filter-feeding sharks known only from a few widely scattered specimens throughout the world. This is a large, unmistakable shark with a flabby body and a huge terminal mouth with numerous short but sharply pointed teeth. Very little is known of its life history other than this epipelagic species exhibits a diurnal vertical migratory pattern following its preferred prey items. Four of the 17 reported specimens have come from California waters.

MEGAMOUTH SHARK FOLLOWS ➤

MEGAMOUTH SHARK   *Megachasma pelagios*

DESCRIPTION: The Megamouth Shark has a large, flabby body, tapering posteriorly; a short, broadly rounded snout; a huge, broadly arched, terminal mouth; large gill openings; and a large, elongated caudal fin, with a long upper lobe and a prominent ventral lobe. The teeth are small, numerous, and hook shaped. Coloration is a dark blue to brownish black or gray above becoming paler on its flanks, and abruptly white below the level of the pectoral and pelvic fins. Posterior fin margins and apices are white. There is a bright white band extending along its upper jaw; the lining of the lower jaw is silvery with dark mottling. Tooth count: 55–108/75–128.* Vertebral count: 139–151.* Spiral valve count: 23–24.*

HABITAT AND RANGE: Megamouth Sharks are warm-temperate to tropical epipelagic sharks found over continental shelves, around oceanic islands, and far offshore in open waters. They have been found from the surface to a depth of at least 166 m over very deep water of up to 4,600 m; a few specimens have washed ashore.

These sharks are known in the eastern Pacific from only four specimens taken off southern California. The first California specimen was taken in a drift gill net set 8 miles off the east end of Catalina Island less than 38 m from the surface in water over 800 m deep. The second was caught in a drift gill net 7 miles west of Dana Point, a location only 20 miles from where the first was caught, in water less than 100 m deep. A third specimen caught in a drift gill net was taken 30 miles off San Diego in water over 350 m deep. A fourth was caught in a drift gill net 42 miles northwest of San Diego in water over 850 m. All four of these specimens were taken during the autumn months of October (three) or November (one). Elsewhere, Megamouth Sharks are known from several widely scattered locations in the Atlantic, North Pacific, and Indian Oceans.

NATURAL HISTORY: Although unconfirmed, the Megamouth Shark is assumed to be oophagous as are other lamnoids. Males mature at 4 to 4.5 m and grow to at least 5.5 m. Females are mature at 5.4 m. The smallest known specimens measured 1.8 and 1.9 m in length. Mating scars have been observed on adult females, but little else is known about their reproductive biology.

Megamouth Sharks are a slow-swimming, filter-feeding species that consume mainly euphausid shrimps, copepods, and jellyfishes. This is one of the few shark species in which daily vertical depth migrations associated with light levels have been confirmed. In a study off southern California using an acoustic transmitter, a single Megamouth was found to maintain an average daytime depth of 120 to 166 m, but at dusk ascended to 12 to 25 m over very deep water.

There are no known predators of the Megamouth. However, craterlike wounds on some specimens appear to have been made by Cookiecutter Sharks. A Megamouth observed swimming at the surface in Indonesian waters was purportedly being attacked by a group of three Sperm Whales.

HUMAN INTERACTIONS: The Megamouth was perhaps one of the most spectacular shark discoveries of the twentieth century. This large, oceanic shark was first recorded in 1976 from off the northern Hawaiian Islands. Since then about 17 specimens have been either caught or photographed, four of which were from off southern California. They are of no commercial value but of considerable value to scientists studying these fascinating animals.

NOMENCLATURE: *Megachasma pelagios* (Taylor, Compagno, and Struhsaker, 1983). The generic name *Megachasma* comes from the Greek *megas,* meaning large, and *chasma,* meaning open mouth. The species name comes from the Greek *pelagios,* meaning open sea, in reference to its pelagic habitat. The common name, Megamouth, refers to its huge mouth.

REFERENCES: Lavenberg and Seigel (1985); Lavenberg (1991); Nelson et al. (1997).

Thresher Sharks (Alopiidae)

The thresher shark family consists of a single genus comprising three species, all of which occur in California waters. The most distinguishable feature of these sharks is the extremely long tail fin, which is about as long as the body trunk. The genus can be subdivided into two distinct groups. One group consists of thresher sharks with a relatively small eye, a thin tail, and no marked grooves on the top of the head; this group includes the Common and Pelagic Thresher Sharks. The second group includes thresher sharks with extremely large eyes, a broad tail, and distinct grooves on the top of the head, running from a central point over the eyes, out and back over the gill region; the sole member of this group is the Bigeye Thresher Shark. Thresher sharks have a worldwide distribution, ranging from temperate to tropical waters. They are found in nearshore coastal waters, including enclosed bays, as well as in oceanic habitats far from land. They have been taken at a depth of 500 m, but most are found within 65 m of the water's surface. All are oophagous, with small litters of two to four pups. They feed on a wide variety of schooling fishes and cephalopods. Thresher sharks use their tails to herd prey species into a tight school and then by rapidly whipping their tails they stun and kill individual fishes or squids. Threshers are the only modern sharks, other than sawsharks (Pristiophoridae), to use a structure other than their jaws and teeth to kill prey. These are one of the most commercially important groups of sharks. A substantial fishery for these sharks developed in the early 1980s in southern California. Despite their great length, half of which consists of their long tail, thresher sharks have a relatively small mouth and teeth and thus are not considered dangerous to humans. In fact, these sharks have often been observed by divers underwater without incident.

1a   Head with a deep horizontal groove extending around each side. Eyes are very large, with orbits expanded onto the dorsal head surface . . . .
. . . . Bigeye Thresher Shark (*Alopias superciliosus*)

1b   Head without a deep horizontal groove extending around each side. Eyes are small, with orbits not expanded onto the dorsal head surface 2

    2a   Flanks above pectoral and pelvic fins dark. Head is narrow and snout is elongated. Labial furrows absent.
    . . . . Pelagic Thresher Shark (*Alopias pelagicus*)

2b Flanks above pectoral and pelvic fins white. Head is broad and snout is short. Labial furrows present . . . . Common Thresher Shark (*Alopias vulpinus*)

PELAGIC THRESHER SHARK   *Alopias pelagicus*

DESCRIPTION: The Pelagic Thresher is the smallest thresher shark with eyes moderately large, but not extending onto the surface of the head; pectoral fins with a nearly straight anterior margin and broadly tipped apices; and a very narrow caudal fin tip. The teeth are small with a single, oblique, 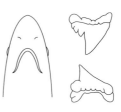 smooth-edged cusp, and except for the first four or five rows all have one or two lateral cusplets; the teeth are similar in both jaws. Color in life is a brilliant dark blue above and white below, which does not extend above the pectoral or pelvic fins. This color rapidly fades to gray after death. Tooth count: 37–43/38–48.* Vertebral count: 453–477.* Spiral valve count: 37–39.*

HABITAT AND RANGE: The Pelagic Thresher Shark is primarily oceanic, although individuals occasionally wander close inshore. They are found near the surface to a depth of at least 152 m. Very little is known about their population structure or movements off the southern California and Mexican coasts. The eastern North Pacific population appears to be centered off southern Baja, but shifts northward during strong El Niño events.

Pelagic Thresher Sharks are usually found off southern California during warm-water years. Because they are often misidentified as the more abundant Common Thresher Shark, they probably have a much wider distribution than is presently known. Elsewhere they are known from scattered records throughout the tropical and warm-temperate Pacific and Indian Oceans.

NATURAL HISTORY: Oophagous, with litters of two pups per birth; one embryo develops per uterus at a time. The average sex ratio at birth is 1:1. The size at birth is 158 to 190 cm. Males mature at 267 to 276 cm with a maximum length of 347 cm. Females mature at 282 to 292 cm and grow to a maximum length of at least 383 cm. Gestation time is unknown for this species. They do not appear to have a defined breeding season as pregnant females have embryos at different developmental stages throughout the year. A single female produces about 40 young throughout her lifetime.

Males mature at seven to eight years and live about 20 years. Females mature at eight to nine years and live about 29 years. The growth rate for juveniles is moderate at 6 to 9 cm per year for the first six years but slows to less than 4 cm per year after the sharks reach maturity.

Pelagic Threshers feed on small schooling fishes and squids, but little else is known about their food habits.

HUMAN INTERACTIONS: Pelagic Threshers are of minor importance to commercial fisheries as they comprise only about 2 percent of the overall number of thresher sharks caught in California waters.

Despite their size these sharks are fairly harmless to people.

NOMENCLATURE: *Alopias pelagicus* (Nakamura, 1935). The generic name *Alopias* is derived from the Greek, meaning foxlike. The specific name *pelagicus* and the common name refer to the oceanic habitat of this species.

REFERENCES: Hanan et al. (1993).

BIGEYE THRESHER SHARK   *Alopias superciliosus*

DESCRIPTION: The Bigeye Thresher is a large thresher shark with extremely large eyes extending onto a dorsal head surface; the head has distinct lateral grooves extending from above the eyes to behind the gill slits (appearing helmetlike); pectoral fins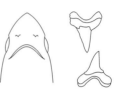
with a curved anterior margin and broadly tipped at the apices; and a broad caudal fin tip. The teeth are moderately large, with a single, long, narrow cusp; they are similar in both jaws. Color is a violet to purplish brown above fading to white below, but with no white patches above the pectoral fin bases. Tooth count: 22–24/20–23. Vertebral count: 301–304. Spiral valve count: 45.*

HABITAT AND RANGE: Bigeye Threshers are usually found over continental shelf waters and in the open sea, occasionally coming close inshore. They are usually caught in areas in which the surface water temperature is 61 to 77 degrees F. They have been caught at a depth of 500 m, but most are found within 65 m of the water's surface.

Bigeye Thresher Sharks are occasionally seen in southern California waters, usually between August and November, and are commonly found in the Gulf of California. Elsewhere the Bigeye Thresher has a worldwide distribution, occurring mostly in warm-temperate and tropical waters.

NATURAL HISTORY: Oophagous, with litters of two to four pups; most have only one embryo per uterus. Males mature at 270 to 288 cm and reach a maximum length of 410 cm. Females mature at 332 to 356 cm and grow to at least 461 cm, with unsubstantiated reports of individuals reaching 5 m. Size at birth is 130 to 140 cm. Based on the best available evidence they do not appear to have a defined breeding season as pregnant females have embryos at various developmental stages year-round.

Male Bigeye Threshers mature in about 9 to 10 years and live at least 19 years. Females attain maturity in about 12 to 14 years and live about 20 years.

Bigeye Threshers feed on benthic and pelagic fishes, squid, and crustaceans. The large eyes are especially adapted for low light levels, and the expanded orbits allow the eyes to roll upward enabling these sharks to hunt by searching for silhouettes of potential prey items above.

As with many other mackerel sharks the Bigeye Thresher can maintain a body temperature several degrees above that of the surrounding water temperature. This may enable it to venture into deeper, cooler water in search of food.

HUMAN INTERACTIONS: Very little is known about their population structure or stock status here or elsewhere. The Bigeye Thresher comprises only about 8 percent of the thresher shark species caught in the California drift gill net fishery. It has been estimated that a single adult female will produce only 20 young during her lifetime, thus making Bigeye Threshers highly susceptible to overfishing.

NOMENCLATURE: *Alopias superciliosus* (Lowe, 1841). The Latin name *superciliosus* refers to the distinct lateral grooves above the enormous eyes of this particular thresher shark. The common name also comes from this shark's very large eyes.

The species was first described in 1839 from Maderia, but was subsequently overlooked by researchers until the 1940s when the scientific name was resurrected based on a detailed description of several specimens from Cuba and Florida. The first confirmed record from California waters was in 1963 and was based on two specimens caught in gill nets three weeks apart.

REFERENCES: Fitch and Craig (1964); Hanan et al. (1993).

COMMON THRESHER SHARK　　*Alopias vulpinus*

DESCRIPTION: The Common Thresher is a large thresher shark with moderately large eyes, but not extending onto the dorsal surface of the head; pectoral fins with a curved anterior margin and pointed apices; and a narrow caudal fin tip. The teeth are relatively small, with a single, narrow, smooth-edged, slightly oblique, triangular cusp; they are similar in both jaws. Color is a silvery to bluish-gray above and white below, with white patches extending above the pectoral and pelvic fin bases. Tooth count: 33–52/31–51. Vertebral count: 359–362. Spiral valve count: 33–34.

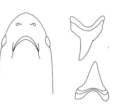

HABITAT AND RANGE: Common Threshers range from shallow inshore waters, including enclosed bays and lagoons, to the open ocean far from land, but they are most abundant within 40 miles of shore. They are usually at or near the surface but have been taken down to 366 m deep. These are strong swimming sharks with a pronounced seasonal migration northward following warm-water masses. Adults, mostly males, move as far north as Vancouver Island in late summer and early fall, but retreat as the water cools. Juveniles tend to remain in the Southern California Bight, which is an important nursery area, rarely venturing further north than Cape Mendocino.

The most common of the three thresher shark species off Baja and southern California, this shark is seasonally abundant northward to Vancouver, British Columbia. It is widely distributed throughout most tropical and temperate seas.

NATURAL HISTORY: Oophagous with litters of two to four. Size at maturity appears to vary worldwide, but for the eastern North Pacific population males mature at about 333 cm and females at about 420 cm. The maximum size for males is 493 cm and for females is 636 cm. Size at birth is 120 to 150 cm. Mating occurs during summer, usually July to August, followed by a gestation period of about nine months. Birth is in spring from March to June. Females may produce only 30 to 40 young during their life span.

Males mature in five years and females in seven years; both sexes live to at least 15 years.

Common Threshers feed mainly on small schooling pelagic fishes, including anchovies, hake, herring, lancetfishes, mackerel, and sardines, as well as on pelagic red crabs and squids.

HUMAN INTERACTIONS: The Common Thresher became the target of an intense commercial fishery beginning in the late 1970s and peaking in 1982 with over 2.3 million pounds being landed. As seen before, the shark fishery grew rapidly and then quickly collapsed as the stock declined from overfishing. The Common Thresher fishery still exists, but more as a by-catch to the lucrative Swordfish fishery. It is believed that the eastern North Pacific population of Common Threshers consists of a single, coastalwide stock with very little immigration from other North Pacific stocks. Although Common Threshers are frequently caught carrying Japanese tuna hooks, indicating long-distance movements, immigration from other populations into the California stock is slow and was not sufficient to replace the number of Common Threshers taken annually at the height of the fishery. As the number of sharks caught declined, the size of individual sharks also decreased. As the fishery collapsed, extensive management regulations were imposed and remain in effect today. The stocks will most likely take years to return to levels that could again support a commercial fishery. Worldwide, thresher sharks are an important target species in many commercial fisheries. The meat is of a high quality for eating.

Common Threshers have small teeth and weak jaws, making them relatively harmless to humans. They have not been implicated in any attacks. Scuba divers encountering them underwater report that the thresher sharks appear to be indifferent to their presence.

NOMENCLATURE: *Alopias vulpinus* (Bonnaterre, 1788). The specific name *vulpinus* is derived from the Latin *vulpes,* meaning fox. Some literature accounts in fact list the species name as *A. vulpes.* Other common local names include Thresher Shark and Whiptail Shark.

Early literature accounts described the Common Thresher as being purplish or bluish gray. The purplish color would seem to be consistent with that of the Bigeye Thresher, indicating that both species were known but not distinguished from each other in California waters until 1964.

REFERENCES: Bedford (1987, 1992); Cailliet and Bedford (1983); Cailliet et al. (1983); Hanan et al. (1993).

Basking Sharks (Cetorhinidae)

The basking shark family is represented by a single wide-ranging species. This enormous shark, exceeded only by the Whale Shark in size, is distinguished by its large fusiform body; short bulbous snout; extremely long gill slits that extend onto the dorsal and ventral head surfaces; a huge gaping mouth with numerous small hooklike teeth; and rough filelike dermal denticles. Basking sharks are found in cold to warm-temperate waters but are apparently absent from equatorial waters. Basking sharks are seasonally abundant along the California coast, often cruising at the surface with their mouth agape filtering plankton from the water. Although very abundant in some areas, little is known about their life history.

BASKING SHARK FOLLOWS ➤

BASKING SHARK

Cetorhinus maximus

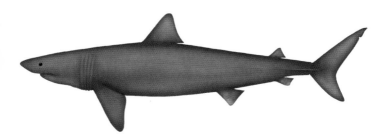

DESCRIPTION: The Basking Shark is a large, stout-bodied shark with a pointed snout; huge mouth; enormous gill slits; strong lateral keels on the caudal peduncle; and a large crescent-shaped tail. The teeth are minute, with a single smooth-edged, hook-shaped cusp. Color is a mottled bluish gray to gray or brown above, becoming variably lighter or darker below.

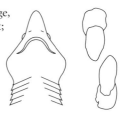

Tooth count: 200+/200+. Vertebral count: 110. Spiral valve count: 47–50.

HABITAT AND RANGE: Basking Sharks are a coastal pelagic species usually found in areas in which the water temperature is between 46 and 75 degrees F. They are found close inshore, including enclosed bays, from the surfline to well offshore at depths of over 500 m, but usually over the continental shelf. Basking Sharks are very social animals and are often seen swimming at the surface in small groups of three to 10, although groups of up to 500 or more individuals have been observed.

Basking Sharks are highly migratory, appearing and then disappearing seasonally at specific localities. Seasonally they are most abundant off central California between October and April. They migrate northward off Washington, British Columbia, and Alaska during spring and summer. They tend to appear in areas of high plankton abundance.

Although common from the Gulf of Alaska to the Gulf of California, Basking Sharks are far more abundant north of Point Conception. Elsewhere, this is a circumglobal species, frequent in temperate to subpolar regions but rare in equatorial waters. These sharks may migrate through tropical areas following thermoclines of cooler, deeper water, but direct evidence for this remains inconclusive.

NATURAL HISTORY: Oophagous, with anecdotal reports of litters ranging from one to six. During mating Basking Sharks swim in a very tight circle, with some swimming snout to tail and some swimming parallel to others, often nudging or biting the pectoral fins of the adjacent, presumably female, shark. Birth size is estimated at 1.5 to 2 m. Juvenile Basking Sharks less than 3 m long are rarely seen. The smallest reported free-swimming Basking Shark measured 1.7 m. The smallest reliably measured specimen from California waters was 3 m long, suggesting that their pupping and nursery grounds are located in plankton-rich oceanic waters at higher latitudes and far away from populated coastal areas. Males mature at 4 to 5 m and females at 8 to 9 m. The maximum size for this species ranges between 12 and 15 m, but most are less than 10 m in length.

Basking Sharks have been estimated to reach maturity in six to seven years, although this has never been validated and may underestimate their age by one-half.

Basking Sharks are near-surface filter feeders consuming vast quantities of zooplankton, such as copepods, the larvae of barnacles, crustaceans, and fish eggs. An individual shark may have up to one-half ton of food in its stomach. Basking Sharks move into nearshore waters during periods of peak plankton abundance to feed, actively foraging to locate areas in which preferred food items, such as swarms of large copepods, are plentiful.

Basking Sharks differ from the other two filter-feeding sharks by relying entirely on the passive flow of water over their gill rakers. An adult Basking Shark cruising at a constant speed of two knots passes about 2,000 tons of water over its gills per hour. The Whale Shark and Megamouth Shark actively pump or gulp water to ingest their prey. Basking Sharks feed by swimming with their mouths open, allowing water to pass over erect gill rakers, which act as a sieve to filter out plankton. They will periodically close their mouths, usually every 30 to 60 seconds, to ingest the filtered plankton. Mucus secreted by glands at the base of the gill rakers traps the plankton as it is filtered.

Basking Sharks periodically shed their gill rakers early in winter when plankton levels are low; replacement takes four to five months. It was hypothesized that Basking Sharks migrate to deeper water, possibly to the edge of the outer continental shelf or into submarine canyons, to "hibernate" on the bottom while their gill rakers regenerate. This seems unlikely as they are usually seen in great number off California at this time. The significance of the loss and replacement of gill rakers by these sharks is uncertain, despite much speculation.

Basking Sharks may migrate considerable distances using oceanic thermal fronts of differing water temperatures as cues or may migrate vertically into deeper water to locate preferred prey. Similar to other mackerel sharks, Basking Sharks are able to maintain their body temperature above that of the surrounding water.

Although there is a seasonal coastal and onshore movement by Basking Sharks, no complete migratory route has yet been established. In California they are most abundant at the surface during autumn and winter but may appear at any time of the year in response to periodic increases in plankton levels. Tagging studies in central California have shown that Basking Sharks will return seasonally to the same area provided a preferred prey species is abundant.

Basking Sharks are generally considered to be sluggish, "basking" at the surface, although at times they have been observed to jump out of the water. The reasons for this behavior are unknown other than it appears to be seasonal and may not be related to feeding.

Adult Basking Sharks have few predators, probably because of their enormous size. However, young specimens are preyed upon by Great White Sharks, Sperm Whales, and Killer Whales.

HUMAN INTERACTIONS: Basking Sharks were historically fished in central California waters for their liver oil, their fins used for soup stock, and as food for human consumption. The fishery first developed in the mid-1920s in central California and continued sporadically into the 1960s. In Monterey Bay in the late 1920s and 1930s local entrepreneurs, usually commercial fishermen, would take customers out to harpoon these sharks for sport. Presently there is no fishery for these sharks in California coastal waters.

Basking Sharks are fairly docile and are easily approached by boats and divers. In fact, they will sometimes actually approach divers in a curious, investigative manner. On occasion divers will hitch a ride on these sharks by grabbing their fins, but caution should be exercised due to the shark's immense size and powerful tail fin, which can cause severe injury.

NOMENCLATURE: *Cetorhinus maximus* (Gunnerus, 1765). The generic name comes from the Greek *cetio,* meaning large or monstrous, and *rhine,* meaning rough or filelike, in reference to its rough skin. The species name comes from the Latin *maximus,* meaning great or large, in reference to its enormous size. The common name refers to its habit of basking at the surface.

REFERENCES: Baduini (1995); Ebert (1992); Phillips (1948); Squire (1967, 1990).

Mackerel Sharks (Lamnidae)

The mackerel sharks consist of three genera and five species of large, active-swimming sharks. All three genera and four of these species occur in California waters. These large, heavy-bodied, fusiform-shaped sharks have a long, conical snout; large mouth with protrusible jaws; large triangular, bladelike teeth; small to moderately large eyes; large gill slits; strong lateral caudal keels; and a short, nearly symmetrical, crescent-shaped caudal fin. All lamnids have a highly developed circulatory system that prevents heat from dissipating through the circulating blood and gills by an elaborate countercurrent heat exchange system. This allows them to elevate their body temperature 7 to 20 degrees F above ambient water temperature. Mackerel sharks are coastal and oceanic, typically found close inshore, including enclosed bays and estuaries, at the surface, and down to depths of 1,280 m. Litters range from two to 16, with the young being born at a fairly large size, which greatly increases the chance of survival from predation. All of the lamnids occupy a niche at or near the top of the food chain. The diet of these sharks is rather broad and varied and includes large bony fishes, cephalopods, cartilaginous fishes, marine reptiles, and marine mammals. Adaptation for high-speed swimming allows them to catch other fast-swimming species such as Swordfishes, tunas, seals, dolphins, other sharks, and rays. All lamnids follow a seasonal migratory pattern related to their main prey species, but very little is known about these movement patterns in California waters.

This group contains some of the largest and most fearsome predatory sharks, including the Great White Shark of popular

media and movie fame. The Great White Shark has been implicated in more attacks on humans and boats than any other species
(the Mako is a close second in number of attacks on boats). The
lack of other large sharks, particularly requiem sharks, in temperate waters and the distinctive appearance of the Great White
Shark make it easier to identify the Great White Shark as the culprit of an attack. Some lamnid species, particularly the Mako and
Salmon Sharks, are the target of major fisheries worldwide. The
Great White Shark, though less common than the other lamnids,
is sought after for its jaws, which are often hung as trophies.

1a Teeth broadly triangular and serrated.....................
    ....Great White Shark (*Carcharodon carcharias*)
1b Teeth long, narrow, and without serrations..............2

    2a Teeth with small lateral cusplets, origin of the first dor-
        sal fin over the base or inner margin of the pectoral
        fins, and a secondary keel on the caudal peduncle ....
        ....Salmon Shark (*Lamna ditropis*)
    2b Teeth without small lateral cusplets, origin of the first
        dorsal fin posterior to rear tip of inner margin, and
        lacking a secondary keel on the caudal peduncle 3

3a Underside of snout and mouth area white; cusps of anterior
    teeth recurved forward; pectoral fins much shorter than
    head length; anal fin originating below the second dorsal fin
    base............Shortfin Mako Shark (*Isurus oxyrinchus*)
3b Underside of snout and mouth area dusky; cusps of ante-
    rior teeth straight; pectoral fins about equal to head length;
    anal fin originating below or posterior to the second dorsal
    fin insertion ....Longfin Mako Shark (*Isurus paucus*)

GREAT WHITE SHARK   *Carcharodon carcharias*

DESCRIPTION: A stout, spindle-shaped body, with a short, conical snout; large, black ominous-looking eyes; a large, high, triangular first dorsal fin and a large crescent-shaped caudal fin characterize the Great White Shark. The teeth are large, triangular, and serrated.  Color is a dusky to dark gray or blue-gray above, becoming abruptly white below. The tips of the pectoral fins are dusky and a black spot may be present on the pectoral fin axil. Tooth count: 24–29/21–24. Vertebral count: 172–185. Spiral valve count: 47–54.

HABITAT AND RANGE: Great White Sharks are found in coastal nearshore areas, including enclosed bays and estuaries, but may at times be oceanic, as individuals are occasionally found around islands far from any mainland. Great White Sharks range from the surface to a depth of 1,280 m, with the average being less than 80 m. Larger Great White Sharks, usually over 3 m, are found in waters with a broader temperature range, from as cool as 45 degrees F to as warm as 80 degrees F. Juvenile Great White Sharks, however, tend to migrate northward, following the movement of warm water masses and occasionally migrating into central and northern California waters during extreme El Niño periods.

In the eastern North Pacific Great White Sharks are found from the Gulf of Alaska to the Gulf of California. They are a cosmopolitan species generally found in cold-temperate to tropical waters, although they are most common in temperate waters between 54 and 68 degrees F.

NATURAL HISTORY: Oophagous, with litters of three to 14 young. The size at birth is 1.2 to 1.5 m. Males mature at about 3.6 m and grow to about 5.5 m. Because few pregnant individuals have been captured and accurately measured, size at maturity for females is problematic, but 4.5 to 5 m appears to be a close approximation. The maximum size of this shark has been variously reported to be 6.4 m, 7.1 m, or 11 m. The largest accurately measured Great White Shark, captured off Cuba in the 1940s, was 6.4 m. The largest reliably measured Great White Shark from California waters was a 5.7-m female. There is an unconfirmed report of a 10-m Great White Shark caught off Soquel Point, California, in the late nineteenth century.

Great White Sharks segregate along the coast by size: juveniles under 1.5 m and adults over 5 m are common south of Point Conception, particularly around the Channel Islands, whereas intermediate and large animals are seasonally abundant in central and northern California waters. The low frequency with which pregnant females have been captured suggests that they may segregate away from the main coastal population and that only a small proportion of the population is gravid at any one time. This may reduce the risk of predation, most likely by other Great White Sharks, on the young and optimize their chance of survival. Warm-temperate areas, such as several of the islands off Baja and southern California, appear to serve as pupping and nursery areas for this species.

Male Great White Sharks mature at about 9 to 10 years and females at about 14 years. A female measuring 6 m was estimated to be 23 years old. Their growth rate has been estimated to be 25 to 30 cm per year for juveniles but slows once they reach maturity. Great White Sharks probably live to a maximum age of about 30 years.

The Great White Shark is one of the most formidable of extant large marine predators, rivaled perhaps only by the Killer Whale. Their large size, powerful jaws, and huge triangular teeth, combined with explosive swimming speed, make them a true superpredator. They feed on a broad spectrum of prey species such as bony fishes, other sharks, rays, and chimaeras. Their prey includes large items such as Basking Sharks and whales as carrion as well as smaller cetaceans such as Dwarf and Pygmy Sperm Whales, White-sided and Risso's Dolphins, and Harbor Porpoises. Sea otters and Northern Fur Seals are frequently wounded or killed by these sharks, but whether they actually consume them is unknown. Great White Sharks have been observed to bite and release Sevengill Sharks. They are known to attack, but not consume, inanimate objects such as boats, kayaks, crab traps, float bags, and buoys. Occasionally found in their diet are sea turtles and birds. Invertebrate prey is rare, but includes abalone, octopus, other marine snails, and crustaceans.

Larger Great White Sharks, over 3 m, prefer marine mammals, whereas small Great White Sharks, less than 2 m, feed more on bony and cartilaginous fishes. Great White Sharks seem to congregate around seal rookeries and other haul out areas, especially when these mammals are breeding. Subadult and young non-breeding adult pinnipeds appear to be most susceptible to predation as they are usually kept out of the breeding areas. Great White Sharks have a voracious appetite. A large female measuring 5.4 m caught off Santa Cruz Island contained two Blue Sharks, a Shortfin Mako Shark, and two sea lions in its stomach. Another large Great White Shark was observed to attack, kill, and consume three elephant seals at the Farallon Islands in a single predation bout.

Great White Sharks may occur either singly or in groups, particularly in feeding aggregations around whale carcasses where nine or more individuals have been observed at a time. A social hierarchy sometimes prevails when feeding on large whale carcasses, with the largest shark feeding first followed in order of size by the other sharks. Blue and other shark species will not feed on a carcass if Great White Sharks are feeding in the area.

Other than humans, the only known predator on Great White Sharks is the Killer Whale, which will feed on them. Great White Shark remains have been found in the diet of Bull Sharks, but how this came about is unknown.

Great White Sharks have often been portrayed as slow and clumsy swimmers, but in fact their swimming is very active and powerful. Tracked with a sonic tag, an individual off the northeastern United States traveled 118 miles in two-and-a-half days with an average cruising speed of 2 miles/hour. Great White Sharks are capable of explosive bursts of speed and in some instances have been known to jump up to 3 m out of the water.

There is good evidence that individual sharks repeatedly return to the same location seasonally year after year. Biologists on the Farallon Islands have been able to identify individual

Great White Sharks based on body markings. The seasonal occurrence of these sharks coincides with peaks in the seal population around these islands.

A distant relative of the Great White Shark, *Carcharodon megalodon*, which once roamed the seas approximately 2 million years ago and is now extinct, has been estimated to have reached a length of 15 m. The only remnants of this species are the enormous fossil teeth and some vertebrae found on fossil beds around the state. Occasional reports of huge Great White Sharks have fueled speculation that these sharks may still be alive; however, these sharks hunted whale species that are now extinct, and it would be unlikely that there would be sufficient prey species left for them to hunt.

HUMAN INTERACTIONS: Although Great White Sharks have never been a target species for fisheries in California, the state nevertheless imposed a ban on fishing for them in 1993. This followed similar bans along the U.S. Atlantic coast, Gulf of Mexico, Australia, and South Africa, where local artisanal fisheries for this species had taken place. Although no demographic studies exist, circumstantial evidence suggests that the number of Great White Sharks in California may be increasing in response to the burgeoning marine mammal population. With California's increasing human population this may inevitably lead to more human–shark interactions.

The Great White Shark is a fearsome predator and is considered to be one of the most dangerous living species of sharks. Media attention has created the image of the Great White Shark as a superpredator of almost mythical proportions. In temperate waters this species has been responsible for more attacks on people, boats, and kayaks than any other shark. In several boat attacks these sharks have repeatedly attacked the boat until it eventually sank. Kayakers have been knocked into the water in some attacks and in one horrific incident two kayakers were killed.

Despite its frightful reputation, the Great White Shark has often been observed in the presence of people swimming, surfing, kayaking, and diving, often coming right up to people and inspecting them before swimming off. Actual attacks by Great White Sharks along the California coast range from a high of seven to a low of zero, with an average of 2.1 attacks on humans per year. Given the millions of people that frequent California beaches yearly, this attack rate is extremely low.

NOMENCLATURE: *Carcharodon carcharias* (Linnaeus, 1758). The generic name *Carcharodon* is derived from the Greek *karcharos,* meaning a sharp point, and *odous* (*odon*) meaning tooth, and is in reference to the huge triangular teeth of this species. The specific name *carcharias* refers to "a kind of shark." The common name comes from the white underbelly of this species. Other local common names include White Shark and Maneater Shark.

REFERENCES: Cailliet et al. (1985); Ebert (1992); Klimley (1985); McCosker (1985); Mollet et al. (1996); Tricas and McCosker (1984).

SHORTFIN MAKO SHARK   *Isurus oxyrinchus*

DESCRIPTION: The Shortfin Mako is a relatively robust, fusiform-shaped shark with a long, acutely pointed snout; eyes less than one-third the snout length; pectoral fin length less than head length; a high first dorsal fin; an anal fin that originates under the midbase of the second dorsal fin; and a crescent-

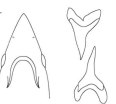

shaped caudal fin. Teeth are similar in both jaws, with a single, large, smooth-edged cusp; four anterior-most upper teeth are somewhat recurved at the tips, with the lateral teeth broader, triangular, and increasingly bladelike. Coloration is a brilliant blue above becoming a lighter blue on the sides to white below. Tooth count: 24–26/22–32. Vertebral count: 183–196. Spiral valve count: 47–48.

HABITAT AND RANGE: The Shortfin Mako Shark is an extremely active-swimming, epipelagic species famed for its jumping ability.

It is found in tropical and warm-temperate waters where the water temperature is above 60 degrees F. Shortfin Mako Sharks follow the movement of warm water masses in which the temperature is usually between 63 and 72 degrees F, moving northward when the water warms but retreating as it cools. They occur from the surface to a depth of about 500 m. Very little is known about their population structure in the eastern North Pacific. Juveniles are fairly abundant in the summer months when they are frequently encountered off Southern California and occasionally as far north as Mendocino. Adults are less common, although they are occasionally encountered on the outer banks of the Southern California Bight, particularly around the Channel Islands during late summer.

Shortfin Makos are found from Oregon southward to Chile. They are most commonly seen off southern California, but move northward seasonally following warm-water masses. Elsewhere, they are a circumglobal species occurring in most temperate and tropical seas.

NATURAL HISTORY: Oophagous, with litters of four to 25 young. Males mature at about 195 cm with a maximum length of 284 cm. Females mature at 270 to 300 cm with a maximum length of 394 cm. Size at birth is about 70 cm. The largest California specimen measured 350 cm, with the average being 210 to 240 cm in length. Shortfin Mako Sharks have a three-year reproductive cycle, which includes a 15- to 18-month gestation period, a late winter to mid-spring pupping season, followed by an 18-month resting period before the female will become fertile again. The Southern California Bight is an important nursery area for juvenile Shortfin Mako Sharks.

Males mature in about four years and females in about seven to eight years. Juvenile Mako Sharks grow quite rapidly, more than doubling their length within the first two years of life. A newborn male weighing 2 to 3 kg will grow to around 140 kg in four to five years whereas a female will grow to around 225 kg in seven to eight years. During the first three years of life Mako Sharks will gain an average of 27 kg/year.

Shortfin Mako Sharks are primarily fish feeders consuming a wide variety of prey species including mackerel, tuna, bonito, anchovies, herring, lancetfishes, rockfish, Lingcod, Yellowtail, sea bass, Swordfish, and juvenile blue, requiem, and hammerhead sharks. Larger Mako Sharks also feed on sea turtles and dolphins, although marine mammal remains are rare in their diet. Invertebrates other than squid are also rare. Estimates indicate that a single mako shark eats over 10 times its weight annually or about 3 percent of its body weight per day.

Shortfin Mako Sharks are known to attack and consume large swordfish. These encounters between large predators often result in injury and even death to the attacking shark. Mako Sharks have been found with swordfish bills imbedded in their body, although on occasion the swordfish may kill the attacking shark. Killer Whales have been observed to attack and consume Shortfin Mako Sharks, some up to 3 m in length.

Shortfin Mako Sharks are able to elevate their body temperature 12 to 18 degrees F above ambient water temperature, which enables these sharks to increase their activity level. Mako Sharks are one of the fastest swimming of all fish species, often running down and overtaking their prey in high-speed attacks.

HUMAN INTERACTIONS: Shortfin Mako Sharks, along with the thresher sharks, were the basis of a large important fishery in southern California in the early 1980s. In 1988 an experimental fishery was established for Mako Sharks in an attempt to develop a sustainable shark fishery. Very little is known about the stock structure of the local Mako Shark population other than juveniles seem to predominate in California's coastal waters. The meat is of high quality and is excellent food.

The Mako Shark is one of the most popular game fish, known for its amazing jumping ability when hooked. During the mid-1980s Shortfin Makos became a popular target species for recreational anglers, with the number of anglers seeking a thrill from catching these sharks growing tenfold.

Attacks on divers and swimmers are relatively rare and few reports are reliable, but these have occurred and suggest that this shark should be regarded as dangerous. The offshore habitat of this species probably prevents it from coming in contact with many swimmers. With its speed, large size, teeth, and aggressive nature it should be treated with extreme caution. A number of attacks on boats have been attributed to this species, most occurring when the shark had been hooked. There are no confirmed attacks in California waters.

NOMENCLATURE: *Isurus oxyrinchus* (Rafinesque, 1810). The generic name *Isurus* derives from the Greek, meaning equal tail, in reference to the crescent-shaped caudal fin with the two lobes being nearly equal. The specific name comes from the Greek *oxys,* meaning sharp or pointed, and *rhynchos,* meaning snout, in reference to its pointed snout. The common name comes from the relatively shorter pectoral fins of this shark compared to its longfinned relative. Other common names include Bonito Shark, Mako Shark, and Mackerel Shark.

The name *Isuropsis glaucus* is used in some early accounts from California. The genus *Isuropsis* was erected by Gill (1862) and the species name *glaucus* was based on a juvenile specimen. Subsequent taxonomic revisions have revealed these other nominal names to be ascribed to *I. oxyrinchus.*

REFERENCES: Cailliet and Bedford (1983); Cailliet et al. (1983); Hanan et al. (1993); Mollet et al. (2000).

LONGFIN MAKO SHARK   *Isurus paucus*

DESCRIPTION: The Longfin Mako is a relatively slender bodied, fusiform-shaped shark with a narrow to bluntly pointed snout; eyes greater than 33 percent of snout length; pectoral fin length equal to head length; a high first dorsal fin; an anal fin that origi-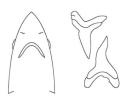nates slightly posterior to the second dorsal fin insertion; and a crescent-shaped caudal fin. Teeth are similar in both jaws, with a single, large, robust, smooth-edged cusp; the four anterior-most upper teeth are not recurved; the lateral teeth are broader, triangular, and increasingly bladelike. Coloration is grayish black above and white below, except for dusky coloration on the underside of the snout and around the mouth, especially prominent in large specimens. Dorsal fins and caudal fin lobes are dusky black; pectoral fins are dusky black above and white below, except for a narrow dusky band along the anterior margin. Tooth count: 26/24. Vertebral count: 195–197.* Spiral valve count: 54.*

HABITAT AND RANGE: The Longfin Mako is an epipelagic species found in tropical and warm-temperate waters off the continental shelf usually at a depth of 120 to 240 m over deeper water. One of the two known California specimens was caught at a depth of 30 m over water 800 m deep.

The only two records for the entire eastern Pacific Ocean came from off Anacapa and Santa Barbara Islands in southern California. Elsewhere, it is known from scattered records in the Atlantic, western and central Pacific, and western Indian Oceans. The Longfin Mako probably has a wider distribution but is frequently misidentified as its better known relative the Shortfin Mako.

NATURAL HISTORY: Oophagous, with litters of two to eight young. The size at birth is between 92 and 97 cm. Maturity is attained at about 190 cm for males and 245 cm for females, with a maximum size of 417 cm. On the Atlantic coast of North America birth takes place during winter.

The Longfin Mako feeds mostly on schooling fishes and pelagic cephalopods.

HUMAN INTERACTIONS: Longfin Makos are of little commercial value as their flesh is mushy and of poor quality. Buyers of seafood in North America will not purchase them.

Although not implicated in any attacks on humans, this species, because of its large size, should be considered potentially dangerous.

NOMENCLATURE: *Isurus paucus* (Guitart, 1966). The Latin for *paucus* means few, which is in reference to the rarity of these sharks relative to their extremely common cousin, the Shortfin Mako. The common name Longfin comes from the extremely long pectoral fins of this shark.

REFERENCES: Ebert (2001).

SALMON SHARK

Lamna ditropis

DESCRIPTION: The Salmon Shark is a stout-bodied, fusiform-shaped shark with a short and conical snout, large eyes, and two caudal keels. Teeth are similar in both jaws, with a single narrow, smooth-edged cusp, flanked by a smaller cusplet on each side. It is dark 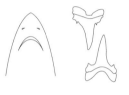 bluish gray to bluish black above, abruptly changing to white below with scattered dark blotches ventrally in adults over 150 cm. The pectoral fins are dark tipped below. Tooth count: 28–32/26–30. Vertebral count: 170–181. Spiral valve count: 37.

HABITAT AND RANGE: The Salmon Shark is a coastal to pelagic species found in subarctic and cold-temperate waters, from the surface to a depth of 375 m, although generally found at greater depths in the southern part of its range. The habitat of Salmon Sharks ranges from close inshore to deep oceanic waters in the north central Pacific Ocean. They are found in water 36 to 75 degrees F but prefer water temperatures between 45 and 66 degrees F. Their movement patterns coincide with seasonal shifts in water temperature. These are swift-swimming sharks able to maintain their body temperature up to 20 degrees F above that of the surrounding water.

Salmon Sharks are encountered from the Gulf of Alaska to central Baja California, becoming progressively more common from central California northward. It is an especially abundant species in the Gulf of Alaska. Elsewhere they are found in the central and western North Pacific Ocean from the Bering Sea to Japan.

NATURAL HISTORY: Oophagous, with litters of two to five young. Size at birth is 65 to 80 cm. Birth usually occurs in spring between March and May after a 12-month gestation. Estimates indicate that a single female will bear about 70 young during her lifetime. Males mature at 180 to 240 cm and females at 194 to 250 cm. The maximum reported size is 305 cm.

Males mature in five years and females in 9 to 10 years. Maximum age is estimated to be 20 to 30 years. Very little is known about their growth rate. A juvenile tagged and subsequently recaptured five months later grew 5 cm during its time at liberty.

A voracious predator on salmon, estimates indicate that the regional population of Salmon Sharks in the central and western Aleutian Islands consumes about 50 million salmon a year. There is no comparable estimate for the eastern Pacific population, but it can be assumed that the consumption of salmon by this shark is considerable. In addition to salmon these sharks feed on pollack, Pacific saury, tomcod, lancetfish, herring, mackerel, spiny dogfish, and squid.

The Salmon Shark is *the* apex predator over much of its range in the North Pacific. Salmon Sharks are known to forage in groups of 30 to 40, using social facilitation to hunt salmon and other schooling species. They follow schools of their main prey species as they migrate around the Pacific Basin. When attacking a school of salmon these sharks usually initiate the attack from below and catch their prey by outrunning it in a high-speed chase rather than ambushing it. They have been observed to chase salmon over long distances and to leap clear of the water with salmon in their mouths.

HUMAN INTERACTIONS: Salmon Sharks are less common than other pelagic sharks in California and are mainly taken as a by-catch to other species. The meat is of high quality and is readily sold along with the fins, which are used for soup stock. In some areas Salmon Sharks are considered an annoyance by fishermen as they destroy fishing gear used to catch more commercially important species. There is virtually no information on their abundance and stock structure in the eastern North Pacific.

Dead or dying young Salmon Sharks often wash up on beaches along the central and southern California coast. The reasons for this are uncertain, but may be related to a rapid warming of the water temperature, which shocks and kills these sharks.

Despite their fearsome appearance Salmon Sharks have not been implicated in any attacks on humans. Divers have observed them underwater without any aggressive or threatening behavior by these sharks. However, they have been known to aggressively approach and bump boats.

NOMENCLATURE: *Lamna ditropis* (Hubbs and Follett, 1947). The scientific name is derived from the Greek *Lamna,* meaning shark, *di,* meaning two, and *tropis,* meaning keel, in reference to the two keels on the caudal peduncle of these sharks. The common name comes from its habit of following its preferred prey, salmon. Other local common names include Mackerel Shark.

Jordan and Gilbert (1880a) first reported this shark in California waters as *Lamna cornubica.* Other records of its occurrence in California waters regarded the Salmon Shark as being conspecific to the Atlantic porbeagle, *L. nasus.* It was not until 1947 that the Salmon Shark was recognized as being a distinct species.

REFERENCES: Croaker (1942); Ebert (1992); Hubbs and Follet (1947).

GROUND SHARKS (CARCHARHINIFORMES)

Worldwide the ground sharks are the dominant, most diverse, and largest shark group, comprising eight families, 49 genera, and at least 225 recognized species. Four families, representing 11 genera and 18 to 20 species, are found in California waters. Some species, such as the Blue Shark, are among the most common sharks in California waters, but many are transient, visiting only when the water temperature warms up during summer months or during extreme El Niño events. The order is characterized by sharks with two spineless dorsal fins, five paired gill slits, a nictitating lower eyelid to protect the eyes, a long mouth extending behind the eyes, and an anal fin. The order contains some of the largest and smallest known shark species, with some exhibiting very specialized morphological adaptations, such as the laterally expanded head of the hammerhead sharks. Most ground sharks are less than 2 m when fully grown, but many of the large, dangerous species, such as the Bull and Tiger Sharks, are members of this order. These sharks occupy a broad range of habitats ranging from cold-temperate to tropical nearshore waters and from water that is shallow to very deep; some are oceanic, whereas others are known to enter freshwater river systems. The reproductive mode of these sharks is variable and may be either egg laying or live bearing, with or without placental viviparity.

1a   Head laterally expanded into cephalofoil or "hammer-shape" . . . . hammerhead sharks (Sphyrnidae)

1b   Head not laterally expanded into cephalofoil or "hammer-shape" 2

     2a   First dorsal fin base originating over or well posterior to the origin of the base of the pelvic fins . . . . catsharks (Scyliorhinidae)

     2b   First dorsal fin base originating well anterior to the origin of the base of the pelvic fins or slightly overlapping them anteriorly 3

3a   Moderate to large sharks, with round eyes, no spiracles, bladelike teeth, short labial furrows, a small second dorsal

fin, precaudal pits, and a scrolled intestinal valve . . . .
. . . . requiem sharks (Carcharhinidae)

3b   Small- to moderate-sized sharks, with oval-shaped eyes,
      large spiracles, teeth pebble- or bladelike, long labial fur-
      rows, without precaudal pits, and a spiral intestinal valve . .
      . . . . houndsharks (Triakidae)

Catsharks (Scyliorhinidae)

The catsharks are the largest family of sharks, comprising 15
genera and at least 106 species worldwide. The number of catsharks in this family will likely increase as several species are
awaiting formal description by taxonomists. Three genera and
four species are currently recognized in California, but this
number will likely increase as another one or two possibly undescribed species have been reported in very deep water off the
California coast. Catsharks can be characterized by their small
slitlike eyes, small multicusped teeth, and a first dorsal fin originating over or behind the pelvic fins. Most of the deepwater catsharks are uniformly colored gray, brown, or black, with few or
no prominent markings. However, many of the nearshore catsharks are among the most colorful shark species with brilliant
patterns of bright spots, blotches, stripes, and prominent, well-
defined saddlebar markings. The catsharks are generally small,
most being less than 80 cm in length, with the largest reaching
about 1.6 m. Catsharks have a vast geographic range from cold-
temperate to tropical waters and are found in all oceans except
the polar regions. The majority are found in deep water along
the outer continental shelves and upper slopes. Most catsharks
are bottom dwellers, at least as adults, and although the juveniles
of some species occupy a midwater habitat, none of them is considered truly pelagic. Most species lay eggs, although some are
known to bear live young. Their diet generally consists of crabs,
squids, octopuses, other invertebrates, small bony fishes, and
other small sharks. In some regions they are an important prey
species for larger sharks.

1a   Ability to swell body trunk into nearly spherical shape by
      swallowing water or air. Body color yellowish-brown above
      with seven or eight saddle bars and numerous dark spots
      scattered over the body surface; the ventral surface is light
      brown. . . . . Swell Shark (*Cephaloscyllium ventriosum*)

1b   Lacks ability to swell body trunk into spherical shape by swallowing water or air. Body color uniform, with no dark spots or saddlebars on back 2

    2a   Anterior margin of upper caudal fin with a prominent row of enlarged dermal denticles . . . . Filetail Catshark (*Parmaturus xaniurus*)

    2b   Anterior margin of upper caudal fin without enlarged dermal denticles 3

3a   Supraorbital sensory canal discontinuous. Upper labial furrows longer than lower. Posterior fin margins not white edged. Anal fin long and low, nearly reaching the base of the caudal fin . . . . Brown Catshark (*Apristurus brunneus*)

3b   Supraorbital sensory canal continuous. Upper labial furrows not longer than lower. Posterior fin margins white edged. Anal fin short and high, not reaching the base of the caudal fin . . . . Longnose Catshark (*Apristurus kampae*)

BROWN CATSHARK   *Apristurus brunneus*

DESCRIPTION: The brown catshark is a small, slender shark with a broadly rounded snout; an internarial space about equal to the nostril length; a discontinuous supraorbital sensory canal; upper labial furrows longer than the lower; relatively small gill openings; two dorsal fins set far back on the body, the first dorsal fin nearly equal to or slightly smaller than the second and originating posterior to the origin of the pelvic fins; a relatively long, low anal fin that extends nearly to the caudal fin base; anal fin insertion posterior to the second dorsal insertion; and an upper caudal fin margin without a crest of enlarged denticles. Teeth are similar in both jaws, with a larger central cusp followed by one or two smaller cusps on each side; the cusps are smooth-edged, and the central cusp in adult males is proportionally higher than in adult females. The coloration is a uniform brown, with the fin edges being slightly darker. Tooth count: 58–74/48–69. Spiral valve count: 14–19. Vertebral count: 116–122.

HABITAT AND RANGE: A poorly known catshark inhabiting the outer continental shelf and upper slopes, the Brown Catshark is usually found over rocky reefs or soft mud bottoms. Juveniles and adolescents occupy a midwater habitat between 200 and 300 m off the bottom and adults occupy a more demersal habitat. They are found at a depth of 33 to 360 m in the northern part of their range and in progressively deeper water, up to 1,000 m, in the southern part of their range. Mesopelagic juveniles and adolescents generally do not inhabit areas in which the water depth is in excess of 1,000 m. Off southern California they have been recorded at depths of 1,298 m. Brown Catsharks prefer areas in which the water temperature is 43 to 48 degrees F.

Brown Catsharks range from southeast Alaska to southern California and northern Baja California. This or a similar species is also found off Central and South America between Panama and Peru.

NATURAL HISTORY: Oviparous, with one fertilized egg case entering each oviduct for a short gestation period prior to being deposited on the bottom, to which it is anchored by its long tendrils. The incubation period in the egg case is long, up to 27 months prior to hatching. There does not appear to be any clear seasonality as to when egg cases are deposited, and they may be deposited year-round. The translucent brown egg case measures about 5 cm in length by about 2.5 cm wide. Two rows of enlarged denticles aid newborns in escaping their egg case at birth. These enlarged denticles disappear shortly after birth. Size at birth is 7 to 9 cm. Males mature at 45 to 50 cm and females at 42 to 48 cm; the largest recorded specimen measured 69 cm.

Brown Catsharks feed mostly on pelagic crustaceans, squids, and small teleosts. Like many other *Apristurus* species, they tend to forage for their food in the midwater column as most of their prey items are pelagic species.

Predators on Brown Catsharks include other sharks and Sperm Whales. Predatory snails feed on the egg cases deposited by these and other egg-laying sharks. The proboscis of these predatory snails is modified such that they can actively drill holes in the egg cases and suck out the protein-rich yolk sac of the developing shark.

HUMAN INTERACTIONS: They are commonly caught in deepwater trawls but are of no commercial value.

NOMENCLATURE: *Apristurus brunneus* (Gilbert, 1892). The genus name is derived from the Greek, *a,* meaning without, *pristis,* meaning file, and (*d*)*urus,* meaning hard. The species name is from the Latin *brunneus,* meaning brown. The common name derives from the uniform brown color of this shark.

The Brown Catshark was originally described as *Catulus brunneus* by Gilbert (1892) from a specimen caught off San Diego. It was subsequently moved to the genus *Scyliorhinus* by Regan (1908) and eventually placed in *Apristurus* by Garman (1913), who erected the genus in his monographic tome on cartilaginous fishes—*The Plagiostoma.*

Apristurus catsharks should be carefully examined as more than a single species of *brunneus*-like catshark may be present off California. This does not include the rare longnose catshark, which may also be represented by more than one *kampae*-like species.

REFERENCES: Cross (1988); Roedel (1951).

LONGNOSE CATSHARK   *Apristurus kampae*

DESCRIPTION: The Longnose is a catshark with a very long snout; a broad internarial space; lower labial furrows longer than the upper; a continuous supraorbital sensory canal; and relatively large gill openings. The first dorsal fin base is about equal in size to

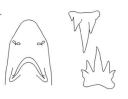

the base of the second dorsal fin, originating over the midpoint of the pelvic fin base; the second dorsal fin is slightly broader and higher than the first dorsal fin, originating slightly behind the midpoint of the anal fin base; the anal fin is relatively short, high, and rounded; its free rear-tip does not reach the origin of the lower caudal fin and its insertion is anterior to the second dorsal fin; the caudal fin is without a crest of enlarged denticles on its upper margin. The color is a uniform blackish or dark brown with the posterior fin edges a prominent white. The inside of the mouth is a grayish-blue. Tooth count: 49–59/ 42–52. Vertebral count: 104–111. Spiral valve count: 9–12.

HABITAT AND RANGE: The Longnose is a deepwater catshark found on the upper continental slope at a depth of 180 to 1,888 m. It generally is found in deeper water than the Brown Catshark, although their depth distribution slightly overlaps.

The Longnose Catshark ranges from Cape Blanco, Oregon, to the Gulf of California. This or a similar species is also found off the Galapagos Islands.

NATURAL HISTORY: Oviparous, with one egg case, measuring 6.8 to 7.1 cm length, deposited per oviduct. Males mature at about 50 cm and females mature at 48 to 52 cm; the maximum length is at least 57 cm. Size at birth is about 14 cm. Newborns are equipped with two rows of enlarged denticles along their back that aid them in escaping from the egg case. These enlarged denticles disappear shortly after birth.

They feed on deepwater shrimps, cephalopods, and small mesopelagic bony fishes.

HUMAN INTERACTIONS: These sharks are occasionally caught in trawls or sablefish traps, but are of no commercial value.

NOMENCLATURE: *Apristurus kampae* (Taylor, 1972). The specific name *kampae* was after Elizabeth Kampa Boden who was the chief scientist on the research cruise that collected the holotype. The common name comes from the relatively long snout of this shark.

An undescribed *Apristurus* species similar to, but distinct from, the Longnose Catshark, also with white fin edges, has been collected from deep water off central California.

REFERENCES: Taylor (1972).

SWELL SHARK   *Cephaloscyllium ventriosum*

DESCRIPTION: The Swell Shark is a small, stout-bodied shark with a short snout, wide mouth, and flattened head. It has the unique ability to distend its stomach by swallowing water or air, swelling its trunk into a nearly spherical shape.

The first dorsal fin originates posterior to the origin of the pelvic fins. The teeth, similar in both jaws, are relatively small and numerous, with a high central cusp and one or two smaller lateral cusps on each side. The color is a yellow-brown above with seven or eight saddlelike bars and numerous dark brown spots randomly distributed over the dorsal body surface and fins; the ventral body surface is light brown. Newborns have a similar pattern, but are lighter in color. Tooth count: 55–69/46–85. Vertebral count: 109–112. Spiral valve count: 10–11.

HABITAT AND RANGE: The Swell Shark is a nearshore bottom-dwelling shark common on rocky reefs, particularly in and around kelp beds and areas with a lush algal-covered bottom. A warm-temperate to subtropical species, they range from close inshore out to a depth of 457 m but are most common at a depth of 5 to 40 m.

The Swell Shark is nocturnal in its habits. During the day it usually lies on the bottom, often in caves and crevices, and sometimes in aggregations piled on top of each other in a group. At night this apparently sluggish shark becomes quite active, swimming through the kelp bed community close to the bottom, foraging for food.

Swell Sharks inhabit an area from Monterey Bay to southern Mexico as well as off central Chile; they are very common south of Point Conception. The absence of recorded observations of this species in the tropical eastern Pacific suggests either that its distribution is not tropical or that few surveys have been conducted in that area.

NATURAL HISTORY: Oviparous, with one encased egg per oviduct at a time. Egg cases are large and purse shaped, with tapering horns and long tendrils. The egg cases are moderately thick walled with smooth surfaces, with some thickening for reinforcement along each lateral margin. Newly laid egg cases, 9 to 13 cm long and 3 to 6 cm wide, are light in color, but darken to an amber-green shortly thereafter, and remain translucent throughout their development. The egg cases from Catalina Island and the mainland population are so strikingly different that they may represent different genetic populations of this species. The gestation period is between 7.5 and 10 months depending on water temperature. Hatching usually occurs at night. The young are 14 to 15 cm at birth. Newborn swell sharks have two long rows of enlarged denticles on their backs that are used to aid them in escaping from the thick-walled egg case. These denticles disappear soon after hatching. Males mature at about 82 to 85 cm, with the maximum recorded size being 110 cm.

Swell Sharks feed mainly on teleosts and crustaceans. The nocturnal pattern of activity of this shark aids it in capturing active-swimming fish, such as blacksmith, that settle down on the bottom and are relatively inactive and unresponsive at night. The egg cases of Swell Sharks are preyed on by several species of boring marine snails and are consumed by certain fish species.

When distressed the Swell Shark has the unique ability to inflate its stomach, similar to a puffer fish, by swallowing water. It will often inflate itself to become wedged in cracks and crevices, making it difficult for a predator to get it out.

HUMAN INTERACTIONS: This is a commercially unimportant species that is occasionally caught by recreational anglers. Its sluggish daytime behavior makes it relatively easy for divers to catch. This is a fairly docile, harmless shark that is often harassed by divers seeking to make the shark inflate itself.

NOMENCLATURE: *Cephaloscyllium ventriosum* (Garman, 1880). The generic name derives from the Greek *kephale*, meaning head, and *skylion*, meaning dogfish; the specific name comes from the Latin *ventricosus*, meaning swell. The common name is in reference to this shark's ability to inflate itself.

The Swell Shark was originally described as *Scyllium ventriosum* by Garman (1880), based on a specimen from Valparaiso, Chile. Jordan and Gilbert (Jordan and Evermann 1896) later described it as *Catulus uter* from a specimen taken off Santa Barbara, considering it distinct from Garman's Chilean Swell Shark. Garman (1913) subsequently combined both into a single species, *Cephaloscyllium ventriosum,* with an antitropical eastern Pacific distribution. Some authors continued to recognize *uter* rather than *ventriosum* until Kato et al. (1967) examined the issue and determined that this was a single species.

REFERENCES: Edwards (1920); Grover (1972a,b, 1974): Nelson and Johnson (1970).

FILETAIL CATSHARK   *Parmaturus xaniurus*

DESCRIPTION: The Filetail is a slender, soft-bodied shark, with a short, broadly rounded snout, and short labial furrows, the lower furrows being longer than the upper. The first dorsal fin is similar in size to the second, originating anterior to the midpoint of pelvic fin bases; the second dorsal fin is smaller than the anal fin. A prominent row of enlarged denticles on the upper caudal fin margin distinguishes it from other catsharks in California. The teeth are similar in both jaws and are small and smooth-edged, with a high central cusp and two to six smaller lateral cusplets. The color is a brownish black above, shading lighter below; the fins have white edges. A light colored band above the pelvic fins extends rearward to the caudal region. The mouth cavity has a white lining. Tooth count: 67–71/78–82. Vertebral count: 109–131. Spiral valve count: 7–8.

HABITAT AND RANGE: A fairly common, but poorly known species, the Filetail Catshark is found on the outer continental shelf and upper slope. Juveniles less than 32 cm in length occupy a midwater habitat, up to 490 m above the bottom in water over 1,000 m deep. With a more demersal lifestyle, adults are usually near the bottom at a depth of 91 to 1,251 m. They are found over rocky and other hard sediment reefs or on muddy, soft bottoms.

Filetail Catsharks have been observed from submersibles at the bottom of the Santa Barbara Basin, a low oxygen area inhabited by few other vertebrates. The enlarged gill slits of this shark appear to enable it to live in areas of low oxygen, such as the Santa Barbara and Santa Cruz Basins and the Monterey Bay Submarine Canyon, where these sharks are very common.

Filetail Catsharks range from Cape Foulweather, Oregon, to the Gulf of California.

NATURAL HISTORY: Oviparous, with two encased eggs per litter. The egg case is long, usually 7 to 11 cm, and slender, about 3 cm wide, with long tendrils and a tan coloration. The egg case can be distinguished from egg cases of other catsharks by the presence of an unusual T-shaped flange running along its lateral edges. Filetail Catsharks breed throughout the year with no defined season. Males mature at 37 to 43 cm and females at 42 to 48 cm, with a maximum size of 61 cm. The size at birth is 7 to 9 cm.

Filetail Catsharks feed mainly on crustaceans and small bony fishes. Many of the prey items are pelagic species, although these sharks have been observed feeding on moribund lanternfishes in the low oxygen Santa Barbara Basin, apparently taking advantage of fishes that had succumbed to the low oxygen environment.

HUMAN INTERACTIONS: These sharks are occasionally taken by trawlers or caught in sablefish traps but are of no commercial value.

NOMENCLATURE: *Parmaturus xaniurus* (Gilbert, 1892). The generic name *Parmaturus* is derived from the Latin *parmatus,* meaning armed with a shield, and the specific name comes from the Greek *xanion,* meaning crest. The common name is in reference to the prominent crest of dermal denticles on the upper caudal fin margin.

Gilbert (1892) originally described the Filetail Catshark from a specimen caught off southern California and placed it in the genus *Catulus.* Garman (1906) subsequently placed *C. xaniurus* in his newly erected genus *Parmaturus,* which included catsharks

that combined a short snout, nostrils close to the mouth, and a crest of enlarged denticles on the upper anterior caudal margin.
REFERENCES: Cross (1988).

Houndsharks (Triakidae)

The houndsharks are a large family consisting of nine genera and 39 species worldwide, of which three genera and five species are found in California waters. These are small to medium-sized sharks, with slender, fusiform bodies; large oval eyes; small to moderately large spiracles; two spineless dorsal fins, the first usually similar in size to the second except in the Soupfin Shark in which the first is much larger; and a weak to strongly developed lower caudal lobe. Maximum size ranges from slightly less than 1 m to nearly 2 m. Along the California coast these sharks generally inhabit shallow coastal waters, bays, and estuaries. Development is viviparous, with or without a placental attachment, with litters between two and 52 young. Their diet consists primarily of crustaceans and teleosts. Historically, some of these species have formed the basis of important commercial and recreational fisheries in California.

1a Well-defined, bold, black saddlebars and dark spots on the dorsal surface. . . . . Leopard Shark (*Triakis semifasciata*)
1b No saddlebars, dark spots, or other prominent markings on the dorsal surface 2

2a Second dorsal fin is much smaller than first dorsal fin and about as large as anal fin, terminal lobe of caudal fin about half the length of the dorsal caudal fin margin . . . . Soupfin Shark (*Galeorhinus galeus*)
2b Second dorsal fin is nearly as large as first dorsal fin and much larger than anal fin, terminal lobe of caudal fin less than half the length of the dorsal caudal fin margin. 3

3a Posterior margin of dorsal fins frayed . . . . Brown Smoothhound Shark (*Mustelus henlei*)
3b Posterior margin of dorsal fins not frayed 4

4a First dorsal fin originating behind pectoral fin, with posterior margin slightly inclined backward, lower lobe of caudal fin indistinct . . . . Gray Smoothhound Shark (*Mustelus californicus*)

4b   First dorsal fin originating over free rear-tips of pectoral fins, with posterior margin vertical from fin apex, lower lobe elongated and pointed in larger specimens . . . .
. . . Sicklefin Smoothhound Shark (*Mustelus lunulatus*)

SOUPFIN SHARK   *Galeorhinus galeus*

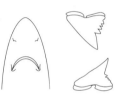

DESCRIPTION: The Soupfin is a large, slender-bodied shark, with a long narrow snout and large mouth; the teeth are small, subtriangular, and blade-like, with a single smooth-edged cusp followed by four or five smaller cusplets; the second dorsal fin is much smaller than the first, but similar in size to the anal fin; the caudal fin has a large subterminal lobe that is nearly as long as the lower lobe. The color is gray to bronze above and white below; the fin edges on juveniles are black. Tooth counts: 32–41/31–46. Vertebral counts: 123–146. Spiral valve counts: 4–5.

HABITAT AND RANGE: The Soupfin is a common coastal shark found in temperate, continental shelf waters from close inshore, including shallow bays, to offshore waters up to 471 m deep and often near the bottom.

Soupfins range from British Columbia to the Pacific coast of central Baja California. Elsewhere they are found in temperate waters of the South Pacific, eastern North Atlantic, South Atlantic, and southwestern Indian Oceans.

NATURAL HISTORY: Viviparous, without a yolk sac placenta, with litters of 6 to 52 young, with 35 being the average. The litter size increases in larger sized females. Males mature at 135 to 150 cm, with a maximum size of 175 cm and females at about 160 cm, with a maximum size of 195 cm. Mating takes place in spring with a gestation period of about 12 months. Pups are born in spring at a size of 30 to 40 cm.

Southern California below Point Conception is an important Soupfin nursery ground, and considerable numbers of adult females and newborn Soupfins are therefore found in this area in spring. San Francisco and Tomales Bays were once important nursery grounds, but since the fisheries of the 1930s and 1940s the number of Soupfins has declined and to this day remains below historical levels.

Coastwide there is a preponderance of adult males in the northern part of the state and females in the south; in central California the sex ratio is about even. South of Point Conception adult males tend to be found in deeper water (at a depth greater than 20 m) whereas females are found closer inshore (at a depth less than 15 m). Soupfins often occur in small schools that are segregated by size and sex.

Males mature in eight to nine years and females in about 11 years. The maximum estimated age for these sharks is about 60 years.

Soupfins are highly migratory, moving to the north during summer and the south during winter or into deeper waters. They are swift moving and can travel up to 34 miles per day and have been reported to travel at a sustained rate at 10 miles per day for up to 100 days. One Soupfin tagged off Ventura was captured 26 months later off Vancouver Island, British Columbia. Another shark tagged in San Francisco Bay was recaptured 12 months later in the same location.

Soupfin Sharks will readily forage on both demersal and pelagic fish species, including sardines, herring, hake, rockfish, mackerel, anchovies, salmon, smelt, Lingcod, midshipmen, barracuda, croakers, opaleye, surfperch, sole, halibut, sculpins, and Sablefish. Larger Soupfins occasionally feed on other cartilaginous fishes including small sharks, stingrays, skates, and chimaeras. Invertebrate prey includes cephalopods, crabs, shrimps, and lobsters. Young sharks tend to feed more heavily on invertebrates than do adults.

Natural predators on the Soupfin, particularly juveniles, are the Great White Shark, Sevengill Shark, and possibly marine mammals.

HUMAN INTERACTIONS: The Soupfin Shark was the mainstay of the shark fishery boom years of 1936 to 1944 when over 24 million pounds were landed. This fishery decimated the Soupfin population, particularly the nursery areas in San Francisco and Tomales Bays. Since 1977 the fishery has roughly averaged between 150,000 and 250,000 pounds annually. These sharks are now mostly taken as a by-catch to other commercial species and by recreational anglers. The meat is of excellent quality for human consumption and the fins are used as soup stock.

The Soupfin has never been implicated in an attack on humans. However, it will snap vigorously when captured and has sufficiently sharp teeth to warrant respect.

NOMENCLATURE: *Galeorhinus galeus* (Linnaeus, 1758). The name derives from the Greek *galeos,* meaning a kind of shark, and *rhinos,* which refers to its long snout. The fins of this shark are highly prized in some cultures for soup stock, hence its common name.

Until recently, most literature accounts of eastern North Pacific Soupfin Sharks listed the species as *Galeorhinus zyopterus* as described by Jordan and Gilbert (1883a). Subsequent taxonomic studies, though, have concluded that the Soupfin is most likely referable to one single wide-ranging species.

REFERENCES: Ripley (1946); Ebert (2001).

GRAY SMOOTHHOUND SHARK   *Mustelus californicus*

DESCRIPTION: The Gray Smoothhound is a houndshark with a short, narrow head; broad internarial space; relatively small eyes; small mouth; teeth blunt and pebblelike; first dorsal fin closer to the pelvic fins than to the pectorals; and rear margin slopes at an angle from the fin apex. The posterior dorsal fin edges are

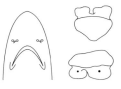

smooth edged. The lower lobe of the caudal fin is not elongated. Coloration is dark gray to brown above, without spots or other prominent markings, and white below. Albinism has been reported for this species from Elkhorn Slough, Monterey Bay. Tooth counts: 70–95/69–94. Vertebral counts: 145–157. Spiral valve counts: 7–8.

HABITAT AND RANGE: This is a common, warm-temperate to sub-tropical, inshore bottom-dwelling shark that is often found in shallow sandy bays. In central California waters it is primarily a winter visitor, but it is a permanent resident in the warmer waters of southern California. It is usually found in water less than 12 m deep but has been caught down to 46 m.

The Gray Smoothhound ranges from Cape Mendocino to Mazatlán, Mexico, including the Gulf of California, although it is rare north of Elkhorn Slough, Monterey Bay; the holotype was reportedly collected in San Francisco Bay. It overlaps the Brown Smoothhound in the northern part of its range and the more tropical Sicklefin Smoothhound in the southern part of its range. All three *Mustelus* species overlap in their distribution between southern California and the Gulf of California.

NATURAL HISTORY: Viviparous, with a yolk sac placenta. Litters range from 3 to 16. Gestation takes about 10 to 12 months. Males mature at 57 to 65 cm and grow to a maximum length of at least 116 cm. Females mature at about 70 cm, with a maximum reported length of 125 cm. The size at birth is 23 to 30 cm.

Males mature at one to two years and live to at least six years. Females mature at two to three years and live to at least nine years. Sharks of both sexes appear to grow rapidly over the first one to three years of life, with their growth rate slowing down at maturity.

Gray Smoothhounds tend to move around in schools and often have been observed to congregate with Leopard Sharks in very shallow water.

These sharks feed mainly on crustaceans, primarily crabs. Smaller sharks tend to feed more on shore crabs, whereas large sharks consume more cancer crabs. Minor dietary items include ghost shrimps, innkeeper worms, and small bony fishes.

HUMAN INTERACTIONS: These sharks are regularly caught by recreational anglers in southern California. They are of little economic value, but in the Gulf of California they are fished commercially by long-liners and utilized as food.

These are small, relatively harmless sharks.

NOMENCLATURE: *Mustelus californicus* (Gill, 1864). The generic name is derived from the Latin *mustela,* meaning weasel, and the specific name *californicus* from the region in which it was first described. Its common name comes from the gray coloration that distinguishes it from the other smoothhound sharks.

REFERENCES: Talent (1982, 1985); Yudin and Cailliet (1990).

BROWN SMOOTHHOUND SHARK *Mustelus henlei*

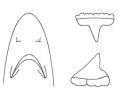

DESCRIPTION: A slender-bodied hound-shark, it is easily distinguished from the other local houndshark species by the frayed posterior margins of the dorsal fins and a long caudal peduncle. The first dorsal fin originates over the pectoral fins. The teeth are strongly cuspidate and sharp. Coloration is brown to bronzy above and white below; it usually lacks spots, although specimens from Humboldt Bay occasionally have minute dark spots present. Tooth counts: 60–80/55–78. Vertebral counts: 152–160. Spiral valve counts: 7–8.

HABITAT AND RANGE: The Brown Smoothhound is an abundant inshore cold to warm-temperate water houndshark typically found swimming near the bottom from the intertidal area out to a depth of at least 200 m. It is the most cool-temperate of the three houndsharks found in California waters, with the other two species found mostly in the warmer waters of southern California. In central and northern California this is one of the most abundant sharks found in shallow Bays and estuaries, particularly in Humboldt, San Francisco, and Tomales bays. Seasonally, during winter, this shark moves from these bays, when the salinity drops due to increased freshwater runoff, to the open ocean.

The Brown Smoothhound ranges from Coos Bay, Oregon, to the Gulf of California. Elsewhere it is found off Ecuador and Peru.

NATURAL HISTORY: Viviparous, with a yolk sac placenta. Litter sizes are from one to 10 with the number of offspring increasing in larger sized adults. Several central and northern California bays are important nursery areas. Brown Smoothhounds give birth in May or June after a 10- to 11-month gestation period. Males mature at 52 to 66 cm and females at 51 to 63 cm. The maximum size for males is 80 cm and for females 100 cm. The size at birth is 19 to 30 cm.

Males mature in about three years and live to at least seven years. Females mature in two to three years and live to about 13 years. Brown Smoothhounds grow quite rapidly during the first few years of life, but growth slows at maturity.

The movement patterns of Brown Smoothhounds are fairly localized, although one tagged individual traveled 96 miles in three months. They move about in discrete groups of sharks of similar age, size, and sex. In San Francisco Bay the sex ratio of this species changes over time and space, with females predominating at times (4.5:1) and males predominating at other times (4:1).

Brown Smoothhounds have been observed to approach gill nets set in shallow water (less than 2 m) at high speed and at the last moment leap over the net. Similarly, Brown Smoothhounds in holding tanks have been observed to swim at high speed at vertical walls, but abruptly change direction 90 degrees and go straight up the wall while maintaining a close proximity to the wall's surface.

The Brown Smoothhound feeds mostly on crustaceans such as crabs, ghost shrimp, and mantis shrimp. Larger individuals occasionally consume small bony fishes including anchovies, perch, smelt, and small flatfish, whereas juveniles will consume fish eggs when available. The type of crustacean species eaten changes with growth as small (less than 60 cm) Brown Smoothhounds consume shore crabs (grapsoids) in higher proportion and larger sharks principally consume cancer crabs.

Brown Smoothhounds are very agile, quick swimmers. They will often swim directly at and then past a crab, swiftly turning 180 degrees, grasp the crab from behind, and quickly crush and ingest it in rapid succession. They may also swim at the crab head-on, grabbing a claw and vigorously shaking it side-to-side to break it off. Once disabled the carapace is crushed, usually from the side or behind, with the entire crab being ingested. These attacks are usually quite rapid from initiation to full consumption. When hunting for food these sharks usually swim close to the bottom and in schools.

Brown Smoothhounds are an important prey item for the Sevengill Shark in Humboldt and San Francisco Bays.

HUMAN INTERACTIONS: Brown Smoothhounds are frequently caught by recreational anglers in San Francisco Bay and elsewhere along the coast. They are not taken commercially in California waters, but in the Gulf of California they are taken by long-liners and utilized as food for human consumption.

These relatively harmless sharks pose little threat to humans.

NOMENCLATURE: *Mustelus henlei* (Gill, 1863). The specific name *henlei* is in honor of F. G. J. Henle, who contributed much to the early classification of elasmobranch systematics. The common name is in reference to the common dorsal background coloration of this species. This shark is also sometimes referred to as "Henle's shark."

The Brown Smoothhound was originally described as *Rhinotriacis henlei* by Gill (1863), who later proposed a new genus, *Pleuracromylon* (Gill, 1864). The problem with Gill's original description was that he compared the Brown Smoothhound with the genus *Triakis* rather than *Mustelus*. Starks (1917), along with other contemporary authors, considered the genus *Rhinotriacis* to be a junior synonym of *Mustelus*.

REFERENCES: Ebert (1986c); Russo (1975); Talent (1982); Yudin and Cailliet (1990).

SICKLEFIN SMOOTHHOUND SHARK   *Mustelus lunulatus*

DESCRIPTION: The Sicklefin Smooth-hound is a houndshark with a short, narrow head; broad internarial space; relatively small eyes; long mouth; and upper labial furrows shorter than the lower. Teeth are blunt and pebblelike.

The first dorsal fin originates over the pectoral fin, with the rear edge abruptly vertical from the apex. The posterior dorsal fin margins are not frayed. The tip of the lower caudal lobe is somewhat pointed and the lower lobe of the caudal fin is elongated, pointed, and hooked. Coloration is gray to olive brown above, lighter below, and lacking any spotting or other distinguishing marks. Tooth counts: 74–102/71–106. Vertebral counts: 133–154. Spiral valve counts: 7–9.

HABITAT AND RANGE: This is a poorly known nearshore, bottom-dwelling houndshark inhabiting warm-temperate and tropical waters in the eastern Pacific.

Sicklefins range from southern California, where they usually appear during warm-water years, particularly near San Diego, to the Gulf of Panama. They are very common in the Gulf of California and southward to Panama.

NATURAL HISTORY: Viviparous, but little else is known about its life history. Males mature at 70 to 83 cm and females at about 97 cm. Size at birth is 32 to 35 cm. The largest recorded specimen measured 110 cm, but they are said to reach 175 cm.

Its diet consists mainly of crustaceans and small teleosts.

HUMAN INTERACTIONS: In California waters they are too uncommon to be of any commercial importance, although they may on occasion be caught by recreational anglers. Further south in the Gulf of California this species is taken commercially for human consumption. It is a nuisance at times to commercial fisheries as large schools often get entangled in gill nets, damaging the gear.

NOMENCLATURE: *Mustelus lunulatus* (Jordan and Gilbert, 1883). The specific name comes from the Latin *lunatus,* meaning sickle shaped, in reference to its posterior fin edges. The common name for the Sicklefin comes from the strongly concave (sicklelike) posterior fin margins.

REFERENCES: Galvan-Magana et al. (1989).

LEOPARD SHARK   *Triakis semifasciata*

DESCRIPTION: The Leopard Shark is a moderately large houndshark with a short, bluntly rounded snout; a broadly arched mouth; teeth with a single, slightly oblique, smooth-edged cusp usually flanked by one or two smaller

cusplets on either side; and an elongated caudal fin with a strong subterminal notch. The body coloration is the most distinctive feature of this shark. The upper portion of the body is a silvery gray to bronzy gray with large, well-defined, bold, black saddlebars and spotting; the underside is lighter and without markings. Adults tend to have more spots and the center of the saddles tend to be lighter colored than in juveniles. Although rare, albinism has been reported. Tooth counts: 41–55/34–45. Vertebral counts: 129–150. Spiral valve counts: 7–8.

HABITAT AND RANGE: This is one of the most common nearshore, cool- to warm-temperate water sharks in California waters. They are usually observed swimming near the bottom at depths of less than 20 m, but have been found at depths of at least 91 m. Leopard Sharks are seasonally abundant in several enclosed bays and estuaries along the coast, including Humboldt, Tomales, Bodega, and San Francisco Bays, as well as Elkhorn Slough, Monterey Bay. It is not unusual to find this shark along the open coast and around the Channel Islands. It may be found over sandy or mud bottoms, in rocky reefs, and in kelp forests.

This species is endemic to the eastern North Pacific, ranging from Oregon to Mazatlán, Mexico, including the Gulf of California.

NATURAL HISTORY: Viviparous, without a yolk sac placenta, with litters of 7 to 36 young produced from an annual reproductive cycle. The gestation period is 10 to 12 months. In central and northern California Leopard Sharks give birth in bays and estuaries during spring, usually in April and May. In Humboldt and San Francisco Bays gravid females release their young within beds of eel grass that not only serve as protection but also provide an abundance of food. Females have also been observed releasing their young in the shallows of Catalina Harbor. Males mature at 70 to 120 cm and reach a maximum length of 150 cm. Females mature at 110 to 130 cm and grow to at least 180 cm, although most are less than 160 cm. There is one record of a female measuring 210 cm. The young are about 20 cm at birth.

Males mature at 7 to 13 years and live to at least 24 years. Females mature at 10 to 15 years and live to at least an age of 20 years. The average growth rate for males is 2.0 cm per year and for females, 2.3 cm per year. Juveniles grow at a faster rate during the first three to four years of life.

The Leopard Shark is an active, strong-swimming shark that forms large schools, sometimes mixing with other species such as Smoothhounds, Spiny Dogfishes, Sevengills, or Bat Rays. Large schools may suddenly appear in a localized area for a time and then just as quickly disappear. These sharks often segregate into schools by size and sex. A school of adult females was once observed feeding on the mud flats in Humboldt Bay while another school, consisting exclusively of adult males, was seen later the same day also feeding on the mud flats, but at a different location. Newborn Leopard Sharks have been observed to move about in a loosely organized group, which not only serves as a means of protection, but may afford the newborns an advantage for hunting as a group can cover a much broader area while searching for food.

Although several tagging studies have been carried out, its movement patterns remain elusive. The Leopard Shark's population structure along the California coast may consist of several regional stocks with limited genetic exchange. Within several central and northern California bays, Leopard Sharks show a strong directional movement pattern that correlates with the tides. During incoming tides they move into shallow mud flats to feed, but they retreat to deeper water as the tide goes out.

Leopard Sharks are opportunistic feeders consuming crustaceans, mollusks, innkeeper worms, and teleosts including anchovies, herring, smelts, croakers, perch, rockfish, flatfish, midshipmen, and, on occasion, young elasmobranchs of other species such as Bat Rays, Spiny Dogfish, and Smoothhound Sharks.

Leopard Sharks have a fairly broad diet that changes with size and season. Larger adults feed mostly on fish, whereas smaller adults and immature sharks consume crustaceans, clam siphons, innkeeper worms, and fish eggs in higher proportion. Smaller juveniles feed mostly on crabs, but frequently consume fish eggs when available during the year. Newborns in Humboldt Bay have been observed feeding on smelt eggs.

Leopard Sharks have been observed to extract clam siphons by running their snout into the mud to grasp the siphon before the clam is able to retract it. The Leopard Shark then contorts its body, thrusting its snout upward, using its body trunk as leverage to pull the siphon, and in some instances the whole clam, out of the mud. Leopard Sharks may use a similar method for extracting innkeeper worms, which also burrow in the mud. Leopard Sharks and Spiny Dogfish have been observed feeding on anchovies in San Francisco Bay by slowly swimming counterclockwise into tightly packed schools of clockwise swimming anchovies. These sharks simply swam into the schools with their mouths agape ingesting the anchovies.

HUMAN INTERACTIONS: Leopard Sharks are mostly taken by recreational anglers but are also taken in a small-scale commercial fishery, particularly in San Francisco Bay. The meat is utilized fresh or frozen and is considered to be of excellent quality.

This shark is generally considered harmless and usually flees when approached by divers. However, there is one record of a diver in Trinidad Bay being harassed by a Leopard Shark.

NOMENCLATURE: *Triakis semifasciata* (Girard, 1854). The generic name comes from the Latin *tri,* meaning three, and *acis* (= *akis*), meaning sharp or pointed, in reference to the tooth cusps. The specific name is derived from the Latin *semi,* meaning half, and *fasciatus,* meaning banded, in reference to the dark saddle markings extending over the dorsal surface on this shark. The common name Leopard Shark comes from its striking color pattern. In early literature accounts this shark was referred to as the Tiger Shark or Catshark.

The name *Triakis californica* was first proposed by Gray (1851), but without a description, thus invalidating the name. The name *Mustelus felis* (Ayres, 1854) was published about one month after Girard (1854) published his description of the Leopard Shark, formally naming it *Triakis semifasciata.*

REFERENCES: Ackerman et al. (2000); Cailliet (1992); Kusher et al. (1992); Russo (1975); Smith and Abramson (1990); Talent (1976, 1985).

Requiem Sharks (Carcharhinidae)

The requiem sharks are one of the larger shark families, comprising 12 genera and more than 50 species worldwide. Four genera and at least six to eight species are known to inhabit California waters. These are what most people picture when they think of a "typical" shark. They are small to very large, with fusiform-shaped bodies; a flattened, but not laterally expanded snout; round eyes with a well-developed nictitating membrane; five paired gill slits, the fifth originating over or behind the pectoral fin origin; bladelike teeth; two spineless dorsal fins, the first much higher than the second and inserting anterior to the pelvic fins; an anal fin; precaudal pits; no caudal keels; a caudal fin with its upper lobe measuring about twice as long as the lower lobe; and a scrolled-type intestinal valve. The color is variable, with most lacking any unusual color pattern. This group dominates the tropical and warm-temperate shark fauna in terms of diversity and number of individuals. These sharks live in a wide variety of habitats ranging from estuaries and enclosed bays, to close inshore on rocky and coral reefs, and out to the open ocean. This is the only shark group that contains species found in freshwater river systems and lakes. The majority of species are found on the continental shelf, but at least two species are found at depths of 400 to 600 m, and three species are exclusively oceanic. Many of the *Carcharhinus* species are very similar and can be difficult to identify. Several of these sharks appear in California waters only during summer and fall months when the water warms or during extreme El Niño events when an influx of unusually warm water attracts many warm-temperate species. In addition to those species confirmed as being in California waters, as many as seven other species usually found off southern and central Baja may show up off southern California during an El Niño year. These include five members of the genus *Carcharhinus*—*C. albimarginatus, C. altimus, C. falciformis, C. galapagensis,* and *C. porosus*—as well as two species of related genera—*Nasolamia velox* and *Negaprion brevirostris.* These species will not be covered further as to date there are no known records of them in California waters. Information on how to identify these species can be found in Compagno et al. (1995). All are live bearing, most with a placental attachment; the litters may be large, but most are between 2 and 20. The requiem sharks are among the most important large marine predators, consuming a broad variety of prey items including bony fishes, sharks, rays, squid, octopus, crabs, sea turtles, sea snakes, and marine mammals. The family contains some of the most dangerous shark species known to humans.

1a   Spiracles present; upper labial furrows very long, extending to front of eyes; prominent lateral ridges on caudal peduncle .... Tiger Shark (*Galeocerdo cuvier*)

1b   Spiracles absent; upper labial furrows not extending to front of eyes; no prominent lateral ridges on caudal peduncle.. 2

2a   Midpoint of first dorsal fin base is closer to pelvic fin origin than to pectoral insertion; color bright blue in life.. .... Blue Shark (*Prionace glauca*)

2b   Midpoint of first dorsal fin base is closer to pectoral insertion than to pelvic fin origin; color gray or bronze in life, but never blue 3

3a   Anal fin origin well in advance of second dorsal fin origin, but never past midpoint of anal fin base; prominent labial furrows .... Pacific Sharpnose Shark (*Rhizoprionodon longurio*)

3b   Anal fin origin under or behind second dorsal fin origin, and always anterior to midpoint of anal fin base; labial furrows not prominent.................................. 4

4a   Interdorsal ridge present 5

4b   Interdorsal ridge absent 6

5a   Dorsal and pectoral fins broadly rounded at apices, with fin tips mottled white or with some black markings, mostly prominent in juveniles . . . .
. . . . Oceanic Whitetip Shark (*Carcharhinus longimanus*)

5b   Dorsal and pectoral fins not broadly rounded at apices, posterior margin of pectoral fins distinctly concave and fin tips not mottled white or black; coloring mostly gray with no distinguishing markings. . . . .
. . . . Dusky Shark (*Carcharhinus obscurus*)

     6a   Fins prominently black tipped . . . .
. . . . Blacktip Shark (*Carcharhinus limbatus*)

     6b   Fins not prominently black tipped. 7

7a   Upper lateral teeth narrow and hook shaped; coloration typically bronze . . . .
. . . . Copper Shark (*Carcharhinus brachyurus*)

7b   Upper lateral teeth broad and triangular; coloration typically gray. . . . .
. . . . Bull Shark (*Carcharhinus leucas*)

COPPER SHARK   *Carcharhinus brachyurus*

DESCRIPTION: The Copper Shark is a large, slender shark, with a moderately long, pointed snout and no interdorsal ridge. The teeth are narrow, triangular, and with a single finely serrated cusp; the uppers are curved and the lowers are erect. The upper teeth are sexually dimorphic as those of adult

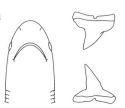

males are longer, narrower, slightly more curved, and with finer serrations than those of adult females and subadults. These sharks are bronze to brownish gray above and white below, with a prominent light band extending anteriorly along its flanks from above the pelvic and anal fins. The fin edges may be dusky, but otherwise have no distinctive markings. Tooth count: 32–36/29–35. Vertebral counts: 196–203.

HABITAT AND RANGE: The Copper Shark is a nearshore species found from the surfline out to 100 m deep or more offshore. The species is migratory, moving into the northern parts of its range during spring and summer, but southward during fall when the water cools. This is the most cool temperate of the *Carcharhinus* species.

The Copper Shark is a seasonal visitor to southern California waters, particularly during summer months. Its eastern Pacific distribution extends from southern California to the Gulf of California and off Peru; it is absent in the tropics. Elsewhere it is a cosmopolitan species in warm-temperate to subtropical seas.

NATURAL HISTORY: Viviparous, with a yolk sac placenta. Litter size ranges from 7 to 20, with an average of 16; those from southern California and Baja tend to have smaller litters. The sex ratio at birth is about 1:1. The size at birth is 59 to 70 cm. Birth occurs in spring and summer after about a 12-month gestation period. Size at maturity may vary between discrete populations, but generally males mature at 200 to 235 cm and females at about 245 cm. Maximum recorded size for males and females is about 300 cm.

Males mature at 13 to 19 years and live to at least 30 years. Females mature at about 19 to 20 years and live to at least 25 years.

Copper Sharks feed on a variety of bony fishes, including demersal and pelagic species such as sardines, smelt, hake, flatfish, cephalopods, and other elasmobranchs. Sharks larger than 2 m tend to feed on elasmobranchs and cephalopods in a slightly higher proportion than smaller specimens.

Copper Sharks often move in migratory groups, their migratory patterns being associated with prey aggregations. They have been observed to hunt cooperatively—"herding" schooling fishes, such as anchovy or sardines, into a tight group, with individual sharks taking turns swimming through the school, mouth agape, engulfing the prey; or capturing larger prey items such as Yellowtail Tuna by swimming in "wing" formation to bunch up the group, with individuals taking turns swimming in and attacking individual tunas. When a particular prey species is in abundance Copper Sharks will feed with other predatory species, such as diving sea birds and other fishes, paying little to no attention to the other predators.

HUMAN INTERACTIONS: The Copper Shark is not abundant enough to be of any commercial importance in California waters. They are caught incidentally by commercial bottom trawlers and long-liners. Smaller specimens are occasionally marketed as food for human consumption. Larger specimens are a popular target species for recreational anglers. Females take approximately 20 years to mature and may live only 25 to 30 years. Taking into account a gestation period of 12 months and a litter size of about 16, the average number of young produced by a female during her lifetime would be about 160 pups. Such a low fecundity combined with the Copper Shark's slow rate of growth, high trophic status, inshore habitat, and ease of capture make this shark extremely vulnerable to overfishing.

Although generally not considered dangerous, the Copper Shark has been implicated in several attacks on humans. Most of these attacks were either provoked or the result of the shark trying to steal speared fish from divers. These sharks will become aggressive, especially toward spear-fishers. There have been no known attacks in California waters.

NOMENCLATURE: *Carcharhinus brachyurus* (Gunther, 1870). The generic name *Carcharhinus* comes from the Greek *karcharias,* meaning a kind of shark. The specific name comes from the Greek *brachys,* meaning short, and *oura,* meaning tail. The Copper Shark derives its common name from its bronze to coppery-colored body. It is also commonly referred to as the Narrowtooth Shark in reference to its tooth shape.

This widely distributed species has been described under several scientific names, having been found in different geographic regions. In California it was described as *C. lamiella* (Jordan and Gilbert, 1883b) based on a juvenile specimen caught off San Diego. Along with the holotype a set of larger jaws apparently referable to the Dusky Shark (*C. obscurus*) was also mentioned in the original description. This complicated the identification of this species and subsequent literature interpretations of *C. lamiella* were referable to *C. obscurus,* which was not distinguished from *C. brachyurus* (= *C. lamiella*) until the mid-1960s. The Copper Shark had also been commonly referred to by the junior synonym *C. remotus* (Dumeril, 1865). Although *C. remotus* is an older name than *C. brachyurus* (Gunther, 1870) the holotype of the former is actually that of another species, the Blacknose Shark, *C. acronotus* (Poey, 1860), a species not even found in the eastern Pacific.

REFERENCES: Rosenblatt and Baldwin (1958); Walter and Ebert (1991).

BULL SHARK   *Carcharhinus leucas*

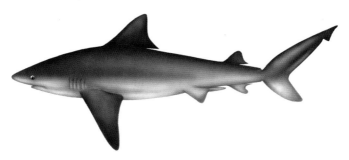

DESCRIPTION: A large, stout-bodied shark, characterized by a very short, blunt snout; a large, broad mouth; and no interdorsal ridge. The upper teeth are broadly tri-angular, slightly curved, and with  a single serrated cusp. The lower teeth are narrower and erect, except for the posterior-most three or four teeth, which are slightly curved. The dorsal surface is gray to grayish brown becoming lighter on the sides to white or yellowish-white below; a pale streak is visible on its sides. The fin tips of juvenile sharks are dusky to black, but fade with growth. Adults lack distinguishing fin marks. Tooth count: 25–29/25–25–27. Vertebral count: 206–215.

HABITAT AND RANGE: A common coastal species usually found around southern Baja and in the Gulf of California, the Bull Shark is occasionally found in estuarine and freshwater river systems, particularly in Central American waters. In California waters it is known from only a few unconfirmed records off southern

California, including Santa Catalina Island, Los Angeles Harbor, and La Jolla Cove. Records of this shark, as with most other *Carcharhinus* species in southern California, have coincided with warm-water years.

Bull Sharks are found in shallow water at a depth of less than 1 m to about 30 m and, on occasion, down to at least 150 m. They are usually found near river mouths and lagoons or in muddy areas, in which the visibility is often reduced, which puts them at a hunting advantage.

Bull Sharks can tolerate both hyper- and hyposaline water conditions. This is the only shark, other than the little-known river sharks (*Glyphis* species), that readily penetrates into fresh-water. They are commonly found in warm-temperate and tropical freshwater river systems throughout the world including several located in Central America. In fact, at one time the Central American population of Bull Sharks in Lake Nicaragua was considered to be a distinct species (*C. nicaraguensis*), as they were thought to be landlocked within the lake. It was only later that ichthyologists determined that these sharks were able to access the sea via rivers draining from the lake. Bull Sharks are known to have traveled over 2,100 miles from the sea up the Amazon River system and one was even caught 1,680 miles up the Mississippi River in Alton, Illinois.

Bull Sharks range from southern California, where they are rare, southward to Ecuador and possibly Peru; they are extremely abundant in the tropical eastern Pacific. Only a few specimens have been caught in California waters, all during periods of extreme warm water. Elsewhere this is a cosmopolitan species found in warm-temperate and tropical seas. *Note*: Several records of the Bull Shark in California waters involved misidentification: For example, the jaws of a 310-cm specimen caught off Anacapa Island in 1942 was long considered to be the first record of the Bull Shark in California waters. Subsequent examination of these jaws revealed them to be those of a Dusky Shark. Although there are no verifiable records of this species off California, I provisionally include it here. It is definitely found in central and northern Baja, and its occurrence in California waters would not be unexpected.

NATURAL HISTORY: Viviparous, with a yolk sac placenta with litters of 1 to 13. Males mature at 200 to 226 cm and grow to a maximum length of 299 cm. Females mature at 210 to 240 cm and grow to at least 324 cm and possibly to 340 cm. The size at birth is 55 to 81 cm. Birth usually occurs in spring and summer after a 10- to 12-month gestation period. Nursery grounds for Bull Sharks tend to be located in estuaries or lagoons.

Males mature at 14 to 15 years and females mature at about 18 years. The maximum age for males is 21 years and for females is 24 years. Growth is slow and varied among individuals: 15 to 20 cm per year for the first five years, 10 cm per year for years 6 to 10, 5 to 7 cm per year for years 11 to 16, and less than 4 to 5 cm per year after maturity is attained.

Bull Sharks prey on other elasmobranchs, including dogfish, bramble, sharpnose, hammerhead, and other requiem sharks, including their own species, as well as guitarfishes, sawfishes, skates, stingrays (stingray spines are commonly found embedded in the mouth of these sharks), Manta Rays, California Butterfly Rays, and Bat Rays. Other prey items include sea turtles, birds, dolphins, cetacean carrion, and land mammals such as antelope, cattle, dogs, cats, rats, and humans. Bony fishes and invertebrates including crabs, shrimps, and cephalopods are more commonly found in juvenile Bull Sharks. The Bull Shark does not consume inanimate objects or garbage as does the Tiger Shark.

Bull Sharks are an important predator on other elasmobranchs in nursery areas. Adolescent Bull Sharks are usually excluded from nursery areas as they will cannibalize juveniles of their own kind and, in fact, are a major predator of young Bull Sharks. Gravid females in nursery areas usually do not feed prior to giving birth.

HUMAN INTERACTIONS: Bull Sharks are rare in California waters and are of no commercial importance. A few are taken in the Mexican shark fishery.

The Bull Shark is considered to be one of the four most dangerous sharks in the world. It may be responsible for more attacks in the tropics than any other shark species. Confusion with other carcharhinid species precludes positive identification in many cases. Extreme caution should be used when confronted by this shark. They appear to be more aggressive in turbid water and more cautious in clear water. No attacks in California waters have been attributed to this species, although several of the "Bull" Sharks caught were in excess of 3 m in length.

These are hardy sharks and keep well in captivity, with some specimens having been maintained for over 15 years.

NOMENCLATURE: *Carcharhinus leucas* (Valenciennes, in Muller and Henle, 1839). The species name comes from the Greek *leukos,* meaning white. The common name, Bull Shark, comes from its pugnacious behavior. It is also referred to as the Gambuso Shark.

The description of the eastern Pacific *C. azureus* (Gilbert and Starks, 1904) was based on three specimens collected at a fish market in Panama. Gilbert and Starks considered it closely related to *C. nicaraguensis* but described it as new based on the distinct freshwater habitat of the aforementioned species. The status and identification of eastern Pacific Bull Sharks were further complicated when several authors misidentified *C. azureus* (= *C. leucas*), confusing it with other *Carcharhinus* species. Garman (1913) referred to *C. azureus* as *C. milberti,* which was later determined to be a junior synonym of the Sandbar Shark (*C. plumbeus*). This led to increased confusion by other authors and further misidentification. For example, Beebe and Tee-Van (1941) considered this species to be closely related to the Sandbar Shark, whereas Rosenblatt and Baldwin (1958) identified *C. azureus* as being referable to the Silvertip Shark (*C. albimarginatus*). The holotype of *C. azureus* is definitely referable to the Bull Shark (*C. leucas*).

The status of this species in California waters is still questionable. Because the Bull Shark might be found in California waters during periods of extreme warm water, it is provisionally cited here.

REFERENCES: Fry and Roedel (1945).

BLACKTIP SHARK FOLLOWS ➤

BLACKTIP SHARK   *Carcharhinus limbatus*

DESCRIPTION: The Blacktip Shark is a stout-bodied shark, with a moderately long, pointed snout and no interdorsal ridge. The teeth are narrow and triangular, with a single broad-based cusp; the edges are finely serrated; the teeth  are similar in both jaws. These sharks are bronzy to a gray-brown above, becoming white below. A pale strip extends along the flanks from the pelvic fins to below the first dorsal fin. The most prominent feature of these sharks is their distinctive black fin tips, except for the anal fin (*usually*) and pelvic fins (*occasionally*), which may lack black tips. Large adults may have light or plain fin tips. Color fades to gray after death or in preservative. Tooth count: 29–32/28–32. Vertebral count: 196–203.

HABITAT AND RANGE: The Blacktip Shark is a common inshore species usually found around river mouths, estuaries, shallow bays, coral reefs, and island lagoons. It is usually found in water less than 30 m deep, but occasionally offshore. These sharks can tolerate reduced salinity conditions found in estuaries and around river mouths, but they do not penetrate into fresh water. Blacktip Sharks migrate by following warm water masses during summer but retreat as the water cools in fall and winter.

There are several literature accounts, but no supporting evidence, of the Blacktip Shark in California waters. The nearest confirmed record is from Ensenada, Baja California, which is about 60 miles south of the California border. It would not be unusual to find these sharks off California during periods of intense El Niño events. Elsewhere this cosmopolitan species is found in most warm-temperate and tropical waters.

NATURAL HISTORY: Viviparous, with a yolk sac placenta and litters of one to 10, with most between four and seven. The gestation is 10 to 12 months with the young being born in late spring and early summer. Females drop their young in site-specific nursery grounds. The young are 38 to 72 cm at birth. Males mature at about 130 to 204 cm and grow to a maximum of 226 to 255 cm. Females mature at 120 to 212 cm and grow to at least 257 cm.

Age at maturity varies regionally, but, in general, males mature in four to six years and females in six to eight years. The maximum age for both sexes is about 10 to 11 years. The growth rate of juveniles is quite rapid at 15 to 24 cm per year, slowing to 7 to 13 cm per year for adolescents and to less than 6 cm per year for adults.

The Blacktip Shark is a voracious fish eater, feeding especially on sardines, herring, anchovies, soles, mackerel, groupers, small sharks, particularly houndsharks and sharpnose sharks, guitarfishes, skates, stingrays, and eagle rays, but also cephalopods and crustaceans. This is a very active-swimming shark that will at times leap out of the water, rotating on its axis several times prior to dropping back. This behavior is believed to be used when feeding on schools of small fishes. These sharks tend to travel and feed in groups.

HUMAN INTERACTIONS: If this shark does indeed occur in California waters it is too rare to be of any importance. It may occasionally be taken as a by-catch to recreational and commercial fisheries. Elsewhere they are commonly taken in fisheries and utilized as food for human consumption. They are the second most important shark species taken in fisheries along the Atlantic coast of North America.

Attacks by this species are few although they will approach divers with speared fish and attempt to steal their catch. They are fairly timid toward humans.

NOMENCLATURE: *Carcharhinus limbatus* (Valenciennes, in Muller and Henel, 1839). The specific name *limbatus* comes from Latin, meaning bordered, in reference to its prominent black markings bordering its fin edges. The common name comes from the prominent black fin tips of this shark.

REFERENCES: Hanan et al. (1993).

OCEANIC WHITETIP SHARK   *Carcharhinus longimanus*

DESCRIPTION: The Oceanic Whitetip Shark is a large, distinctive requiem shark, with a heavy body; a short blunt snout; a large, high first dorsal fin; an interdorsal ridge; and long, broad pec-toral fins. The upper teeth are broadly  triangular, slightly curved, and serrated. The lower teeth are nar-rower, erect, and serrated. These gray to brownish colored sharks have very distinctive white mottling present on the first dorsal, pectoral, pelvic, and caudal fins. The underside of the pectoral and pelvic fins are often heavily mottled with brown. Young specimens less than 130 cm have black blotches on their fin tips, but this fades with growth and is replaced by the prominent white mottling. Tooth count: 28–32/27–31. Vertebral count: 232–244.*

HABITAT AND RANGE: The Oceanic Whitetip Shark is an epipelagic species usually found far offshore in the open waters of the tropics, generally where the bottom depth is over 200 m. They range from the surface to a depth of about 150 m, occasionally venturing in-shore in waters as shallow as 37 m, particularly around islands or in areas in which the continental shelf is narrow. They prefer water temperatures over 68 degrees F and will migrate seasonally into areas as the water temperatures rise but withdraw as it declines.

A rare species in California waters, this epipelagic species is occasionally seen around the Channel Islands during warm-water years, with unconfirmed reports of individuals off central California. Oceanic Whitetip Shark are most common between

20 degrees N and 20 degrees S latitude, but will move north and south beyond their normal range following the seasonal movements of warm water masses. A cosmopolitan species, this is one of the most common oceanic sharks in tropical and warm-temperate seas.

NATURAL HISTORY: Viviparous, with a yolk sac placenta with litters between 1 and 15, averaging six. Larger females tend to have larger litters. Gestation is 9 to 12 months. Birth varies somewhat regionally but occurs mainly in spring and summer; the sex ratio at birth is usually 1:1. The size at birth is 55 to 77 cm. Males mature at 175 to 198 cm, with a maximum length of 245 cm. Females mature at 180 to 190 cm and may reach 350 cm, but individuals over 3 m are exceptional. Maturity occurs between four and five years for both males and females. The rapid growth and early maturity appear to play important roles in the survival strategy of this species. The population dynamics of this species is poorly known. They do appear to segregate regionally by sex and size, with males predominating in some locations and females in others.

Oceanic Whitetip Sharks appear to be sluggish, slow-swimming sharks, but they are known to feed on a wide variety of active, fast-swimming prey species including tunas, barracuda, White Marlin, dolphinfish, lancetfish, oarfish, threadfish, Swordfish, and Pelagic Stingray. They also feed on squid, sea turtles, sea birds, and mammalian carrion.

The coloration of these sharks plays an important role in allowing them to closely approach active-swimming prey species. At a distance the gray body coloration becomes shaded out and the white spots on the fin tips appear to emulate schooling fishes. A predator mistaking these spots for schooling fishes will rapidly charge at the suspected "prey items" only to find too late that it has now become the prey of the Oceanic Whitetip Shark. After luring fast-swimming species close to it, an Oceanic Whitetip Shark puts on a burst of speed and quickly overtakes its victim. Oceanic Whitetip Sharks at times will form aggregations and have been observed to feed cooperatively by "herding" schooling fishes and squids into a tight ball and then swimming through it with mouths agape ingesting the prey items. Oceanic Whitetip Sharks are equally active when foraging for food by day or at night.

HUMAN INTERACTIONS: Oceanic Whitetip Sharks are known to cause considerable damage to pelagic long-line fishing gear used for Swordfish and tunas. The meat is marketed for human consumption in some regions and the fins are often used for shark fin soup. They are too rare in California waters to be of any commercial importance.

Their heavy build, strong jaws, large teeth, aggressive behavior, and opportunistic feeding habits make the Oceanic Whitetip Sharks one of the world's four most dangerous shark species. It has been implicated in numerous attacks on survivors of sea disasters. An inquisitive shark, it will readily approach divers working in offshore waters and can quickly become quite aggressive. Very rare in California waters, it is unlikely to be encountered by most shore divers, surfers, and swimmers. No attacks have been attributed to this species in California.

NOMENCLATURE: *Carcharhinus longimanus* (Poey, 1861). The species name comes from the Latin *longus,* meaning long, and *manus,* meaning hand, in reference to its long pectoral fins. The common name is in reference to its oceanic habitat and prominent white fins.

REFERENCES: Backus et al. (1956); Myrberg (1991); Seki et al. (1998).

DUSKY SHARK *Carcharhinus obscurus*

DESCRIPTION: The Dusky Shark is large and slender bodied, with a moderately long, broadly rounded snout, and an interdorsal ridge. The upper teeth are broadly triangular, slightly curved, and serrated. The lower teeth are

narrower, with a single, finely serrated, erect cusp. In life this shark is gray to bronze-gray above and white below, with an inconspicuous light strip extending anteriorly along the flanks from the pelvic area. The fin tips are dusky in juveniles, becoming indistinct in adults, but otherwise with no prominent markings. Tooth count: 29–33/27–33. Vertebral count: 173–184.

HABITAT AND RANGE: Dusky Sharks usually are found on the continental and insular shelves from the nearshore to a depth of 400 m. This shark is highly migratory as it moves northward into southern California waters as the temperature increases during summer but retreats south as the water cools down. They tend to prefer water temperatures between 66 and 73 degrees F. Dusky Sharks avoid estuaries and areas of low salinity.

Dusky Sharks range from southern California to the Gulf of California. This is a cosmopolitan species of warm-temperate and tropical seas. In areas in which they are found, Dusky Sharks may be extremely abundant, particularly juveniles in nearshore waters and over the continental shelf.

NATURAL HISTORY: Viviparous, with a yolk sac placenta with litters ranging from 3 to 14 young. The sex ratio at birth is approximately 1:1. Unlike some *Carcharhinus* species there does not appear to be a correlation between maternal size and litter size. The pupping season may last over several months, usually from late winter to summer. In some areas births may occur year-round with no defined season. Gestation lasts about 16 months, with females not mating for about a year after giving birth. Females move inshore to give birth, and then leave the nursery area usually moving offshore into deeper water. The nursery areas occupied by newborns are distinctive from the areas inhabited by the rest of the adult population, with the juveniles closer inshore. Several lagoons on the Pacific coast of Baja are important nursery areas.

Males mature at about 2.8 m and females at 2.6 to 3.0 m. Males grow to at least 3.4 m and females to at least 3.7 m and possibly to 4.2 m. The young at birth are 70 to 100 cm.

The age at maturity varies regionally. Male Dusky Sharks reach maturity at 19 to 21 years and females at 17 to 24 years. Dusky Sharks live to at least 34 years, with estimates of up to 50 years for some of the larger individuals.

Dusky Sharks eat a wide variety of demersal and pelagic bony fishes, including sardines, herring, anchovies, moray eels, cuskeels, mullet, barracuda, groupers, croakers, jacks, mackerel, tunas, flatfishes, angel sharks, spiny dogfishes, catsharks, smoothhounds, and other requiem sharks, skates, rays, guitarfishes, and various invertebrates including crabs, lobsters, shrimps, octopus, squid, and sea stars. On occasion they will also consume cetaceans as carrion.

Young Dusky Sharks move in large aggregations in search of food. They frequently migrate following the movement of warm water masses in pursuit of preferred prey items. Young Dusky Sharks are an important prey item for several species of large sharks, such as the Great White, Bull, and Tiger Sharks, helping to regulate their populations.

HUMAN INTERACTIONS: Dusky Sharks are an important fishery in some locations, such as Western Australian, where up to 530 tons are landed annually. The fishery is based mostly on newborns less than 100 cm in length. In addition to its meat, which is used as food for human consumption, the fins are used for soup, the hide for its leather, and its liver oil is extracted for vitamins. They are relatively uncommon and of no importance in the California shark fishery.

The Dusky Shark is considered a dangerous species due to its large size and teeth. They have been implicated in several attacks on humans, but none have occurred in California waters. Very little is known about their behavior around swimmers or divers.

NOMENCLATURE: *Carcharhinus obscurus* (Le Sueur, 1818). The species name, *obscurus,* comes from the Latin meaning dark or dim. The common name refers to its dusky body coloration. The Dusky Shark is also referred to as the Bay Shark in some literature accounts.

Although early accounts refer to this species as *C. lamiella,* the holotype is identifiable as *C. brachyurus* and not *C. obscurus.* It appears that Jordan and Gilbert (1883b) may have had the Dusky Shark in mind when describing *C. lamiella,* as they refer to a pair of jaws from a larger but different specimen in their original description. Regardless, the designated holotype of *C. lamiella* is *C. brachyurus* and therefore is synonymized under that name rather than under *C. obscurus.*

REFERENCES: Rosenblatt and Baldwin (1958); Natanson et al. (1995).

TIGER SHARK

Galeocerdo cuvier

DESCRIPTION: The Tiger Shark is a giant, heavy-bodied shark, with a broad head, a very short, blunt snout, a large, broad mouth, long upper labial furrows, and spiracles. The teeth are a large, distinctive cockscomb shape, with heavy serrations. Tiger Sharks are gray above and lighter below, with a striking pattern of black spots and vertical bars on its flanks. These spots and vertical bars are quite vibrant in juveniles less than 1.5 m, but become faint or indistinct in adults and are absent in specimens larger than 3 m. Tooth count: 20–25/21–25. Vertebral count: 216–234.*

HABITAT AND RANGE: Tiger Sharks are a common, wide-ranging species found on continental shelves and insular slopes from the intertidal area out to at least 350 m. They are known to make open ocean excursions far offshore, often between islands, but are not considered oceanic like the Blue Shark or the Oceanic Whitetip. Most common in the tropics, they extend into warm-temperate areas. They move northward as the water warms, but retreat as it cools.

Tiger Sharks are nocturnally active, moving close inshore at night, often into bays or lagoons in which the water is barely deep enough for them to swim, but retreating into deeper water at daylight. Smaller Tiger Sharks appear to be more active in shallow water during daylight than are larger ones, but at night all sizes readily move inshore. A telemetric tag attached to one large (approximately 400 cm) individual showed that it was slightly more active in its swimming behavior during the day, covering a greater area in deeper water, whereas it was less active at night covering less area when it came into shallower water.

Occasionally observed off southern California, usually during warm-water years, and southward, the Tiger Shark becomes increasingly abundant in the tropics, to at least Peru. The few verified California records have coincided with El Niño events. Elsewhere it is a cosmopolitan species found in tropical and warm-temperate seas.

NATURAL HISTORY: Viviparous, without a yolk sac placenta, this is the only ovoviviparous carcharhinid shark. Litters of 10 to 82 pups are produced after a gestation period of 12 to 16 months. Mating takes place during spring, with birth the following spring or summer. Males mature at 226 to 290 cm and reach a maximum size of 381 cm. Females mature at 250 to 350 cm and grow to at least 5.5 m. The size at birth is 51 to 76 cm.

Tiger Sharks are perhaps the largest carcharhinoid species, rivaled only by the Great Hammerhead, *Sphyrna mokarran,* and they are one of the 10 largest species of living sharks. One gigantic female caught in southeast Asian waters was reported to be 7.4 m and to weigh 3,110 kg. The largest Tiger Shark caught in California waters measured 2.7 m.

The age at maturity varies slightly between populations, but in general males mature at seven to 10 years and females at eight to 10 years. The growth of young Tiger Sharks is quite rapid, as they will nearly double their size over the first year of life. Young Tiger Sharks will grow about 20 cm per year until they reach maturity, after which their growth rate slows to 10 cm per year. Males live to at least 15 years and females to 16 years and possibly as long as 20 to 25 years.

Some Tiger Sharks remain in a localized area for some time. However, tagging studies have shown that Tiger Sharks are capable of moving long distances over short periods of time and that home ranges for this species may be quite large. One individual is reported to have traveled over 3,330 miles in a six-month period, and another traveled 1,015 miles in 17 days, a daily average of 58 miles.

The Tiger Shark is legendary for its omnivorous, opportunistic feeding habits. It is one of the few sharks that is a true scavenger, eating a wide variety of prey items including crabs, lobsters, horseshoe crabs, gastropods, cephalopods, jellyfishes, numerous bony fishes, other elasmobranchs, sea turtles, sea snakes, marine birds, and marine mammals. In addition, they readily feed on carrion and have been found with various terrestrial animal remains, including from rats, pigs, cattle, sheep, donkeys, dogs, hyenas, monkeys, and humans. An incredible array of garbage from human origin has been found in these sharks, including leather, fabrics, coal, wood, plastic bags, burlap sacks, cans, license plates, assorted pieces of metal, and other seemingly inedible objects. The Tiger Shark has at times been called "a garbage can with fins!"

The diet of the Tiger Shark changes with growth, with sharks tending to feed heavily on bony fishes and cephalopods when they are small, but less so as they increase in size. Large Tiger Sharks consume mostly elasmobranchs, sea turtles, mammals, sea birds, crustaceans, and indigestible materials. Tiger Sharks over 230 cm begin to feed on prey items similar in size to humans and thus present a greater threat.

Given their large size, adult Tiger Sharks probably have few, if any, predators. On the other hand young sharks, with their slender, flexible bodies and inefficient anguilliform swimming motion, are highly vulnerable to predation, particularly by other sharks, including their own kind. To reduce predation by their own species young Tiger Sharks tend to segregate themselves in coastal nursery areas away from larger Tiger Sharks.

HUMAN INTERACTIONS: The Tiger Shark is too rare in California waters to be of any commercial importance. In areas in which Tiger Sharks are abundant their meat is utilized for human consumption. The fins may be used for soup stock and its liver has a high vitamin A potency when processed for vitamin oil. Its skin is of high quality and is used for leather and other products. It is one of the seven sharks for which sport fishing records are kept. The current all-tackle record of 568 kg was caught on a 60-kg test line.

The Tiger Shark is one of the most dangerous sharks in the world, second only to the Great White Shark, with more confirmed attacks on swimmers, divers, and boats in the tropics than any other species. An attack in California waters on a swimmer in La Jolla Cove in 1959 may have been by a Tiger Shark, but more likely it was a Great White Shark, which has accounted for most of the attacks in California coastal waters. Because the victim's body was never recovered it remains an open question. However, it should be noted that 1959 was an exceptionally warm El Niño year, the kind that often brings many warmer water species, such as the Tiger Shark, northward into California waters.

NOMENCLATURE: *Galeocerdo cuvier* (Peron and Le Sueur, 1822). The generic name comes from the Greek *galeos,* meaning shark, and *kerdaleos,* meaning wily, crafty, cunning, or shrewd. The specific name *cuvier* is in honor of the early nineteenth-century naturalist Georges Cuvier.

Literature accounts cite several nominal names for the Tiger Shark, including *G. tigrinus* by Starks (1917) and *G. arcticus* by Walford (1935) in earlier publications on California's elasmobranch fauna. Most of the supposed differences between the various nominal species appear to be related to descriptions based on specimens of different sizes and from different geographic regions. For example, the descriptions of *G. tigrinus* (Faber, 1829) and *G. arcticus* (Muller and Henle, 1839) were based on small juvenile or embryonic specimens for the former and on an adolescent or adult specimens for the latter. Only a single wide-ranging species is recognized for this genus.

REFERENCES: Scofield (1941); Seigel et al. (1995).

BLUE SHARK   *Prionace glauca*

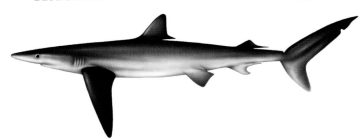

DESCRIPTION: The Blue Shark has a slender body, a long and narrow snout, large eyes, no interdorsal ridge, and long, narrow pectoral fins. The upper teeth have a single, slender, narrowly triangular cusp, with finely serrated edges; the lower teeth are similar, except the cusps are more erect and the edges 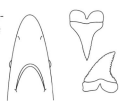 may be smooth or finely serrated. The body color ranges from a dark to indigo blue above, shading to a bright blue on its flanks to white below. This brilliant blue color fades after death to purple or gray. Tooth count: 27–30/27–30. Vertebral count: 239–252.

HABITAT AND RANGE: Blue Sharks are a wide-ranging pelagic, oceanic species found in both temperate and tropical seas from warm equatorial waters up to 81 degrees F to colder waters with temperatures as low as 45 degrees F. They prefer water temperatures between 50 and 68 degrees F. They are found at greater depths (80 to 220 m) near equatorial waters and shallower depths the farther they are from the equator. Although primarily an offshore species, Blue Sharks will often venture close inshore in areas in which the continental shelf is narrow or around offshore islands.

Blue Sharks are common along the entire California coast and, among sharks, are perhaps one of the most frequently encountered sharks. They are often seen slowly cruising at the surface with their first dorsal fin and upper caudal fin cutting through the water's surface and their large pectoral fins splayed out. As the water warms during summer these sharks move northward, extending all the way to the Gulf of Alaska. As the water cools during winter months they retreat to the south. Along the coast, adult females, followed closely by juveniles of both sexes, start their northward migration in early spring. Adult males begin their migration later and tend to stay farther offshore. Between May and October immature sharks are particularly abundant within the Southern California Bight and in Monterey Bay.

The Blue Shark is one of the most wide-ranging of all elasmobranch species, extending throughout the eastern North Pacific from as far north as the Gulf of Alaska, where it is a summer migrant, to as far south as southern Chile.

NATURAL HISTORY: Viviparous, with a yolk sac placenta. Litters may vary from 4 to 135 young, but most average 20 to 40. Mating occurs between late spring and early winter followed by a 9- to 12-month gestation period. The Southern California Bight is a major pupping and nursery area, with Blue Sharks born in the early spring ranging from 34 to 53 cm. Pregnant females have on average a sex ratio for embryos of about 1:1. Males and females both mature at about 220 cm and attain a maximum size of about 400 cm, with unconfirmed reports of up to 600 cm.

Female Blue Sharks become sexually active a year or so before maturing and mate during this time. These females are able to store and retain sperm until the ovaries mature the following year at which time they become fertilized. The pups will be born the following year. Adult females typically bear extensive scars from biting by males during courtship. As an adaptation to this behavior the skin of adult females is about three times as thick as males'.

Males mature in four to five years and females in five to six years, with a maximum age of 20 years. Growth is quite rapid the first few years of life, at least 21 cm per year, but slows down after maturity.

Blue Sharks feed mainly on small prey items, particularly bony fishes and squid, as well as on pelagic crustaceans. Most prey are schooling pelagic species, but bottom-dwelling fishes and invertebrates are also readily eaten. Some of the fish species preyed upon include herring, sardines, anchovies, blacksmith, salmon, lancetfishes, flying fishes, pipefishes, hake, rock cod, mackerel, tunas, sea bass, flatfishes, Spiny Dogfishes, other Blue Sharks, and, in one unusual instance, a Goblin Shark. Pelagic cephalopods, in particular squid, are an important prey species. Blue Sharks will congregate around dead or dying cetaceans and pinnipeds to feed, although they do not actively forage on these as live prey items. Great White Sharks will actively prey on Blue Sharks. When feeding on cetacean carrion, such as a large whale carcass, a Blue Shark will usually give way or risk becoming a meal itself when Great White Sharks show up in the area to feed.

Blue Sharks show several distinct feeding patterns when attacking prey such as anchovies and squids, each depending on the size and activity level of the shark and on the activity and behavior of the prey species involved. For example, when squids form large breeding aggregations Blue Sharks will often appear to feed on them by simply swimming through the spawning mass with their mouths open and ingesting the squid. At other times they will charge through the school, quickly engulfing large numbers of squid. The behavior of the Blue Shark is dependent on the behavior pattern of the squid.

Blue Sharks will also feed in cooperation with other species when a prey item is particularly abundant. In the Southern California Bight Blue Sharks have been observed to feed cooperatively with bonito, barracuda, and diving seabirds when anchovies are abundant. The bonito and barracuda will herd the anchovies into a tight school with the bonito attacking the school from the side, the barracuda from below, and diving seabirds from the surface. Once these other predators have herded the anchovies together the Blue Sharks simply take advantage of the situation by opportunistically swimming through the school, mouths agape, and taking a swathe of anchovies into their mouths. Although this cooperative behavior would seem to benefit the Blue Sharks most, it should be kept in mind that under different circumstances these sharks will actively feed on the bonito and barracuda.

The California Blue Shark population fluctuates seasonally with subadults, usually between one and three years old, most abundant in coastal waters from early spring to early winter. Mature adults are uncommon close inshore. Blue Sharks are known to undergo extensive migrations, in some cases moving over 3,604 miles. One specimen tagged off southern California was recaptured near Midway Island in the central Pacific Ocean.

HUMAN INTERACTIONS: Blue Sharks have periodically been the target of directed commercial fishing efforts in California, but the poor quality of the meat and lack of a market have limited its development. Currently there is no directed commercial fishery in California for Blue Sharks, although they are taken incidentally as a by-catch in the drift gill net fishery off southern California. Recreational fishing, on the other hand, has grown tremendously since 1981 with over 400,000 anglers annually fishing for these and Mako Sharks from charter or private boats. Most of the recreational fishing takes place in southern California. Elsewhere the fins of this shark are retained by long-liners in the central Pacific where they are used for soup stock. The hide from these sharks is used for leather and the meat is utilized for human consumption in some areas, although it is considered of inferior quality to other shark species.

Blue Sharks are considered dangerous and have been implicated in attacks on both humans and boats. There are several confirmed attacks in California waters. Blue Sharks will readily approach divers with speared fish and attempt to steal their catch. The Blue Shark is usually not overly aggressively toward humans but will readily attack if baited-in by thrill-seekers looking for an adrenaline rush or if provoked, as was the case for an amateur scuba diver who attempted to hand-feed a school of Blue Sharks off Santa Barbara Island. Extreme caution should be exercised by divers and swimmers when these sharks are present.

NOMENCLATURE: *Prionace glauca* (Linnaeus, 1758). The generic name comes from the Greek *prion,* meaning saw, and *akis,* meaning point. The specific name *glauca* is from the Latin meaning blue. The common name is derived from the brilliant blue coloration of this shark. It has also been referred to as the Great Blue Shark.

REFERENCES: Cailliet and Bedford (1983); Hanan et al. (1993); Harvey (1989); Sciarrotta and Nelson (1977); Tricas (1979).

PACIFIC SHARPNOSE SHARK   *Rhizoprionodon longurio*

DESCRIPTION: The Pacific Sharpnose is a small, slender-bodied requiem shark, with a long, pointed snout, and, except for the first dorsal fin, relatively small fins. The first dorsal fin is located closer to the pectoral fins than the pelvic fins. This is the only requiem shark of the

eastern Pacific with long labial furrows and the second dorsal fin originating behind the anal fin. The teeth are small with oblique, smooth-edged cusps, except in adults, in whom the upper teeth may be finely serrated. Coloration may be brown or grayish above and white below. The dorsal fins have dusky tips and the pectorals have lighter colored edges. Juveniles have light-colored fin edges, except for the caudal fin margin, which is darker. Tooth count: 27–31/26–28. Vertebral count: 146–167.

HABITAT AND RANGE: The Pacific Sharpnose is a very poorly known coastal requiem shark found from close inshore out to a depth of 27 m.

It is a common shark species in the warm-temperate and tropical eastern Pacific extending from southern California to Peru. It is rare in California waters, usually seen during El Niño years, but is very common around southern Baja California and in the Gulf of California.

NATURAL HISTORY: Viviparous, with a yolk sac placenta, and litters of two to five. Males mature at 58 to 69 cm and grow to at least 92 cm. Females mature by 103 cm and grow to at least 110 cm and possibly to 154 cm. The size at birth is 30 to 34 cm.

Its diet consists mainly of small bony fishes and crustaceans.

HUMAN INTERACTIONS: It is too rare in California waters to be of any importance, but in the Gulf of California, where it is common, the Pacific Sharpnose is fished by long-line and utilized as food for human consumption.

The Pacific Sharpnose Shark is a small, relatively harmless species.

NOMENCLATURE: *Rhizoprionodon longurio* (Jordan and Gilbert, 1883). The generic name comes from the Greek *rhizion,* meaning root, and *prionodon,* meaning saw-tooth. The species name comes from the Latin *longurius* meaning a long pole or rodlike. The common name is in reference to the Pacific Ocean, where it resides, and its sharply pointed snout. This shark is also simply referred to as a Sharpnose Shark.

The Pacific Sharpnose Shark was originally described as *Scoliodon longurio* by Jordan and Gilbert (1883c), but subsequent taxonomic studies placed this species in the genus *Rhizoprionodon.*

REFERENCES: Hubbs and McHugh (1950); Roedel (1950); Springer (1964).

Hammerhead Sharks (Sphyrnidae)

The hammerhead sharks are an unmistakable family that cannot be confused with any other shark group due to their unique "hammer" or "bonnet"-shaped head. The hammerheads are a small carcharhinid group comprising two genera and eight species worldwide, of which three species are known from California waters. These mostly coastal sharks are found close inshore out to a depth of about 275 m. One or more species are found in most temperate and tropical seas. The hammerheads are small to very large sharks with one species growing no more than 1 m in length and another reaching about 6 m, although most are between 1.5 and 4 m. All are viviparous with a yolk sac placenta, and produce litters of 4 to 50 young. Hammerheads feed on a wide spectrum of prey items including bony fishes, other elasmobranchs, squid, octopus, crabs, shrimps, and sea snails. Members of this family have been implicated in attacks on people, although their reputation appears to be greatly exaggerated, as most are fairly timid around humans. Caution should be exercised around them as some of these sharks are quite large and attacks have been verified.

1a   Head expanded into a distinctive bonnet shape . . . .
. . . . Bonnethead Shark (*Sphyrna tiburo*)
1b   Head expanded into a distinctive hammer shape. 2
    2a   Anterior margin of head with a prominent central indentation . . . .
. . . . . . Scalloped Hammerhead Shark (*Sphyrna lewini*)
    2b   Anterior margin of head without a prominent central indentation. . . . .
. . . . . . Smooth Hammerhead Shark (*Sphyrna zygaena*)

SCALLOPED HAMMERHEAD SHARK   *Sphyrna lewini*

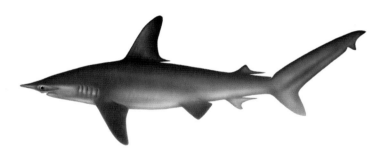

DESCRIPTION: The distinctive feature that sets this hammerhead shark apart from the other two species found in California waters is the "scallop-shaped" indentations on the anterior margin of the head. The teeth of both jaws are generally smooth edged, although in larger sharks they are weakly serrated; the upper teeth have a single, narrow, triangular-shaped cusp; the lower teeth are similar to the upper teeth but are narrower and more erect. Color is light to brownish gray or bronze to olive above and white below. The pectoral fin tips are dusky in adults. The pectoral, second dorsal fin tip, and lower caudal fin tips of juveniles have a dark blotch, but this fades later in life. Albinism has been reported in this species. Tooth count: 30–36/30–35. Vertebral count: 174–206.*

HABITAT AND RANGE: The Scalloped Hammerhead is a coastal species usually found over continental and insular shelves, from close inshore out to a depth of 275 m. Young sharks tend to stay closer inshore, including in bays and other protected areas that function as nursery grounds. Juveniles tend to prefer areas of turbid water conditions. This is a warm-temperate to tropical species usually found in areas in which the water temperature is above 72 degrees F.

Scalloped Hammerheads are a highly mobile, migratory species that at times forms large schools. In the Gulf of California these large schools of sharks congregate around seamounts during the day but, as tracking studies have revealed, disperse at night to feed. During summer months they will migrate to higher latitudes following warmer water, occasionally straying into California waters but retreating once the water begins to cool.

Scalloped Hammerheads are extremely abundant in the Gulf of California but are rare in California waters, with the few records of this species having come during or just after El Niño events. The Scalloped Hammerhead has long been confused with the cool-temperate water preferring Smooth Hammerhead, the most commonly seen hammerhead shark in California. The earliest reported record of the Scalloped Hammerhead was of a 240-cm specimen caught off San Pedro in 1940, an El Niño year; although because the Scalloped Hammerhead is not normally found in water that is less than 72 degrees F this record was later disputed. The first verified record came in 1977, another El Niño year, from off Santa Barbara and was based on the capture of three specimens. Since then several additional specimens have been reported, with these captures usually coinciding with warm-water years.

Because it is difficult to differentiate hammerhead shark species, the Scalloped Hammerhead may frequent California waters more often than indicated. This would be especially true during extreme warm-water years. The Great Hammerhead (*S. mokarran*), which can be distinguished from other hammerheads by its serrated teeth and strongly curved posterior pelvic fin margin, has not been reported in these waters, but it would not be unexpected for a vagrant individual to show up. Elsewhere the Scalloped Hammerhead is a circumglobal species usually found in warm-temperate to tropical seas.

NATURAL HISTORY: Viviparous, with a yolk sac placenta. Litters have 13 to 31 young. Males mature at 140 to 165 cm and grow to at least 295 cm. Females mature at 200 to 250 cm and grow to at least 370 cm, but may exceed 400 cm. Size at birth is 38 to 56 cm following a gestation period of approximately 9 to 10 months. Breeding takes place year-round, with no defined season.

Scalloped Hammerheads segregate during different stages of their life cycle by size and sex. Adult females move inshore to mate and give birth but then move offshore. Newborns remain in the nursery area for several months before moving out into open coastal waters where they become vulnerable to predation by larger sharks, including members of their own kind.

The few confirmed records of this species in California waters include both adult females and newborn pups less than one month old that were captured during years of extreme El Niño events. The capture of Scalloped Hammerheads at these times indicates that this displaced species will readily use southern California as a pupping or nursery area when suitable oceanographic conditions prevail.

Scalloped Hammerheads grow quite rapidly, averaging 10 to 15 cm per year during the first five years of life, but slowing to 5 to 10 cm per year afterward. Because juveniles are preyed upon by larger sharks, their rapid growth allows them to reach a size at which they are less vulnerable to predation. Females mature at about 15 years and males at about 10 years. Scalloped Hammerheads may live up to 35 years.

Scalloped Hammerheads feed mostly on teleosts, including sardines, herring, anchovies, mullet, barracuda, Pacific bonito, mackerel, flatfishes, other sharks including pups of their own kind, and stingrays. Invertebrate prey items include octopuses, squids, shrimps, crabs, and lobsters. Adult and larger subadult males are major predators on young Scalloped Hammerhead pups. Adult females feeding offshore tend to consume more pelagic prey items, particularly squid. Scalloped Hammerheads will often forage in groups when hunting for food.

HUMAN INTERACTIONS: The Scalloped Hammerhead is too rare in California to be of any commercial value. The few confirmed specimens were in bottom gill nets or caught by anglers. In the Gulf of California, where they are abundant, they are taken by local fishers and the meat is utilized as food for human consumption. Fisheries in other regions utilize the meat for food, the fins for soup stock, the skin for leather, and the liver oil for vitamins.

A potentially dangerous species given their size, hammerheads have been implicated in attacks on humans, but as is often the case the identification of the species involved is undetermined. The Scalloped Hammerhead along with the great and Smooth Hammerheads are all large sharks that are easily confused. Recorded attacks on humans as well as on a few boats can usually be attributed only to a "hammerhead shark" without any further identification.

Observations by divers swimming directly into schools of Scalloped Hammerheads have shown that they are relatively timid, fleeing from approaching divers when scuba gear is used. Compared to other species such as the Bull, Tiger, and Great White Sharks, these are far less dangerous.

NOMENCLATURE: *Sphyrna lewini* (Griffith and Smith, 1834). The generic name *Sphyrna* comes from the Latin meaning hammer, and the specific name *lewini* was in honor of John William Lewin (1770–1819), a natural history painter and coroner in New South Wales. The species is sometimes wrongly spelled *leeuwini*. The common name Scalloped Hammerhead comes from the prominent central indentation flanked by the less-defined lateral indentations on each side.

REFERENCES: Fusaro and Anderson (1980); Gilbert (1967); Seigel (1985).

BONNETHEAD SHARK FOLLOWS ➤

BONNETHEAD SHARK

Sphyrna tiburo

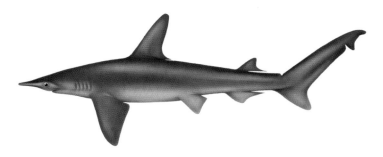

DESCRIPTION: The Bonnethead is the smallest of the three hammerhead shark species found in California waters. This small hammerhead can be easily identified by its unique bonnet- or shovel-shaped head. The anterior margin of the head is evenly rounded, smooth, and without any indentations. The upper teeth are triangular, curved (except for the first tooth, which is erect), and with a single, smooth-edged cusp. The lower teeth are shorter than the upper and with narrower cusps; the first to third teeth are erect, the fourth through seventh or eighth are slightly curved, and subsequent teeth are molarlike, lacking cusps, and are modified for crushing. Color is gray to gray-brown above and white below, occasionally with small dark spots on the side, but otherwise with no conspicuous markings. Tooth count: 25–28/ 25–27. Vertebral count: 163–173.

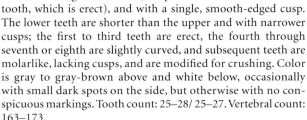

HABITAT AND RANGE: The Bonnethead Shark is a common inshore species found in bays and estuaries in warm-temperate and tropical waters. It is usually found over mud or sandy bottoms and around coral reefs from the surfline to a depth of 25 m and occasionally 80 m.

The Bonnethead ranges from San Diego to northern Peru. It is very common in the Gulf of California but is seen in California waters only during warm-water years. Elsewhere it can be found along the Atlantic coast from Rhode Island, where it is rare, southward to Brazil, including the Gulf of Mexico.

NATURAL HISTORY: Viviparous, with a yolk sac placenta. Litter size is between 4 and 21 young, with an average of nine. Mating takes place in spring with the young being born after a four- to five-month gestation period. The size at birth is 24 to 30 cm. Males mature at 68 to 80 cm and grow to about 124 cm. Females mature at 80 to 90 cm and grow to at least 117 cm and possibly as large as 150 cm.

Males mature in two years and females in about two-and-a-half years. The growth rate is about 20 cm per year for both sexes during the first year of life. Males between the first and second year grow about 10 cm, in the third year about 5 cm, and very little after that. The maximum age for males is six to eight years. Females, on the other hand, grow throughout their lives, but with progressively slower growth each successive year: in years one to two they grow about 13 cm, year three 10 cm, year four 8 cm, year five 6 cm, and after age six about 4 cm per year. Females live at least seven years and possibly up to 12 years in some populations.

Bonnetheads are usually found in schools numbering anywhere from a few individuals up to several hundred and sometimes several thousand. They tend to school by size and sex. During warm-water periods they migrate northward into California waters, but retreat when the water temperature drops. Studies on the behavior of these sharks have revealed a complex pattern associated with a social hierarchy in which some individuals are more dominant. Part of this hierarchy is based on size and sex.

Bonnetheads feed mostly on crabs and shrimps but also on barnacles, bivalves, and cephalopods. Bonnetheads are an important prey item for larger coastal sharks.

HUMAN INTERACTIONS: They are occasionally caught by recreational anglers in southern California but are of little importance. A common inshore species in Mexican waters, they are caught on longlines and in trawls and are utilized as food for human consumption.

There has been one reported attack by this species from an area outside of California. Otherwise, this is a small, generally harmless shark. These rather hardy sharks are frequently exhibited in public aquariums where they usually do quite well, often living for years.

NOMENCLATURE: *Sphyrna tiburo* (Linnaeus, 1758). The specific name *tiburo* comes from Spanish, meaning shark. The common name comes from this shark's unique bonnet-shaped head. Earlier publications referred to it as the Shovelhead Shark.

The Bonnethead of the Pacific was first described as a different species, *S. vespertina*, from the Bonnethead of the Atlantic, *S. tiburo*. Based on subsequent research the two forms were later considered subspecies, the Pacific Bonnethead as *S. tiburo vespertina* and the Atlantic Bonnethead as *S. tiburo tiburo*. Current systematic classification considers both forms a single species.

REFERENCES: Gilbert (1967); Springer (1940).

SMOOTH HAMMERHEAD SHARK   *Sphyrna zygaena*

DESCRIPTION: The Smooth Hammerhead is a large hammerhead shark with a "smooth" anterior head margin, lacking any central indentations or notches. The upper teeth have a single, triangular, oblique cusp; the edges are smooth in the young but are weakly serrated in larger sharks; the lower teeth are similar to the upper but are smaller; the first four teeth are erect, with subsequent teeth curved. Color is a dark olive to brownish gray above, the shading becoming lighter on the sides to white below. The ventral surface of the pectoral fins is dusky to black; there are no other conspicuous markings. Tooth count: 26–32/25–30.* Vertebral count: 193–206.*

HABITAT AND RANGE: This species is a large, active-swimming Hammerhead usually found close inshore, in bays and estuaries, but is occasionally pelagic, as they are found around offshore islands. Adults are found further offshore at a depth of about

200 m near the edge of the continental shelf. They are usually observed swimming at or near the surface to a depth of 20 m. The most cool-water tolerant of the hammerhead shark species, they are rarely found in tropical waters.

Smooth Hammerheads have been observed off southern California, occasionally extending northward into central California during warm-water summers. Their range extends southward into the Gulf of California where they, along with the Scalloped Hammerhead, are both quite abundant. This is the most common hammerhead shark in California waters.

NATURAL HISTORY: Viviparous, with a yolk sac placenta. Litters of 20 to 50 young are born in late spring and early summer after a 10- to 11-month gestation period. The young are 50 to 61 cm at birth. Maturity is reached at about 250 cm for males and 265 cm for females. The maximum size is 350 to 400 cm. Despite its abundance in some areas very little is known about its general biology, and virtually nothing is known about its biology in California waters.

Smooth Hammerheads have been observed in large schools numbering upward into the thousands. They move in a south-to-north migratory pattern following warm water masses.

Smooth Hammerheads feed on a variety of cephalopods and bony fishes, with larger individuals consuming elasmobranchs including smaller sharks and stingrays. Crustaceans are taken to a lesser extent. Large adults feeding offshore prey on pelagic cephalopods.

HUMAN INTERACTIONS: The Smooth Hammerhead is regarded as dangerous and at least one attack, most probably by this species, has been reported in California waters. Off southern California Hammerhead Sharks have stolen catches from recreational anglers and spearfishers. Most attacks by hammerhead sharks in temperate waters can be attributed to this species.

NOMENCLATURE: *Sphyrna zygaena* (Linnaeus, 1758). The specific name *zygaena* is from the Latin, meaning a kind of shark. The common name is in reference to the contour of the anterior margin of its head.

REFERENCES: Gilbert (1967).

RAYS (RAJIFORMES)

Worldwide the rays, or batoids, are the largest shark group comprising 22 families, 71 genera, and at least 543 recognized species, a number most likely to increase to over 600 as new species are described. Ten families, representing 12 genera and 22 species, occur in California waters. Some species are quite common in California waters, while others are transient visitors occurring only when the water temperature warms up during summer months or during extreme El Niño events.

The guitarfishes (Rhinobatidae) and thornback rays (Platyrhinidae), each being represented by a single family in California waters, can be characterized by a broadly rounded or elongate, slightly depressed preoral snout; two subtriangular, moderate-sized dorsal fins, with the first originating well posterior to the pelvic fins; a fairly stout, muscular, precaudal tail; and a caudal fin with a strong upper lobe, but lacking a lower lobe. These are small- to moderate-sized rays with most between 50 and 200 cm in length. They are usually found on continental shelves and upper slopes in warm-temperate to tropical waters. Their mode of reproduction is viviparous without yolk sac placenta. Depending on the species they feed mainly on benthic invertebrates and small bony fishes. In some regions they are commercially important as food for human consumption.

Torpedo rays (Torpedinidae) can be characterized by their enlarged, oval-shaped disc with prominent kidney-shaped electric organs that can deliver a shock of up to 80 volts. The disc surface is smooth without any prickles, spines, or thorns. The pelvic fins are divided into distinct anterior and posterior lobes with their origin being anterior to the free rear-tips of the pectoral fins. The precaudal tail is stout and muscular, without any electric organs or stinging spines. The caudal fin is more or less heterocercal shaped, with or without a lower lobe. These are small to moderate-sized rays ranging from 17 to 180 cm in length. They are found in cool-temperate to tropical seas around the world, mostly in the tropics, usually in shallow water, but on occasion at depths down to 1,071 m. They are viviparous without a yolk sac placenta. They feed on crustaceans, cephalopods, polychaete worms, and fishes. They are of minimal use for human consumption, but are used in biological and medical research.

Worldwide the skates (Arhynchobatidae and Rajidae) represent about one-fourth of all living cartilaginous fish species. In California waters two families and at least 11 species are known to occur. Members of this group are characterized by a rhomboid-shaped disc; a long, broadly angular or rounded preoral snout; five paired gill slits; no anal fin; a slender tail tapering posteriorly, usually with one or more rows of enlarged thorns or smaller thornlets, but no stinging spine; two small dorsal fins set well behind the pelvic fins; and a small caudal fin that may be reduced or absent in some species. Adults of many species show a high degree of sexual dimorphism with the anterior disc margins of males becoming "bell-shaped" compared with that of adult females. Males of some species develop specialized thorns on the dorsal surface of their disc that are used during copulation; alar thorns occur along the lateral disc edges and malar thorns along the anterior disc margin. Skates may range in size from 25 cm to more than 200 cm in length, although the majority are less than 1 m. They range from the polar regions to the tropics, although they are found at much greater depths in more equatorial regions. Most species have relatively narrow ranges as endemism is quite high within this group. At least one species is known only from the Gulf of California. All are benthic species found close inshore, in bays, to a depth of nearly 3,000 m. Skates are generally harmless although their thorns are quite sharp and they can inflict a painful wound if mishandled. All skates are oviparous.

Stingrays are most abundant in tropical to warm-temperate waters with only a very few species ranging into cool-temperate areas. Five families (Dasyatididae, Gymnuridae, Mobulidae, Myliobatidae, and Urolophidae) composed of seven species occur in California waters. Some stingrays are relatively common inshore inhabitants of California's coastal waters, whereas others are only occasional visitors during periods of exceptionally warm water. Although stingrays are more diverse in the warmer waters of Baja, some of these species, such as the Longtail Stingray (*Dasyatis longus*), Pacific Cownose Ray (*Rhinoptera steinachneri*), Roughskin Bull Ray (*Pteromylaeus asperrimus*), Spotted Eagle Ray (*Aetobatus narinari*), and Longnose Eagle Ray (*Myliobatis longirostris*) might occasionally range into California waters during El Niño years. Size will range from stingrays with a maximum disc width of 20 cm to the enormous Manta Ray with a wing span of over 6 m. (Disc width is used because the whiplike tail of these stingrays makes total length a less accurate measurement.) Their mode of development is viviparous, without a yolk sac placenta, but with a system whereby uterine "milk" is secreted to the embryos from the mother through villilike filaments called trophonemata. Stingrays feed on a wide variety of invertebrates and bony fishes. Although of minimal commercial value in California waters, elsewhere some species are of considerable importance to local fisheries.

1a   Caudal fin prominent 2
1b   Caudal fin small or absent 4
    2a   Electric organs present . . . .
         . . . . torpedo rays (Torpedinidae)
    2b   Electric organs absent 3
3a   Pectoral fins and head forming wedge-shaped disc; snout broad to narrowly angular; small thorns around eyes and along midline of back; first dorsal fin closer to pelvic fins than to caudal fin . . . . guitarfishes (Rhinobatidae)
3b   Pectoral fins and head forming heart-shaped disc; snout broadly rounded; one to three rows of large hooklike thorns on disc; first dorsal fin closer to caudal fin than to pelvic fins . . . . thornback rays (Platyrhinidae)
    4a   Tail base with no stinging spine, two dorsal fins . . . . . .
         5
    4b   Tail base with stinging spine, zero to one dorsal fins . .
         6
5a   Snout flabby and flexible . . . .
     . . . . softnose skates (Arhynchobatidae)
5b   Snout rigid and stiff . . . . hardnose skates (Rajidae)
    6a   Caudal fin well developed . . . .
         . . . . round stingrays (Urolophidae)
    6b   Caudal fin absent 7
7a   Head elevated above disc; eyes and spiracles on sides of head 8
7b   Head not elevated above disc; eyes and spiracles on top of head 9
    8a   Head without paired hornlike flaps . . . .
         . . . . eagle rays (Myliobatidae)
    8b   Head with paired hornlike flaps . . . .
         . . . . devil rays (Mobulidae)

9a   Disc width only slightly greater than length; tail long, whip-like, greater than disc length; stinging spine long . . . .
. . . . whiptail stingrays (Dasyatidae)
9b   Disc width nearly twice length; tail short, slender, less than disc length; stinging spine short . . . .
. . . . butterfly rays (Gymnuridae)

Guitarfishes (Rhinobatidae)

The guitarfishes consist of four genera and at least 50 species. Two genera and two species are known from California waters. These rays are characterized by a narrow, angular snout, large angular to rounded pectoral fins forming a wedge-shaped disc with the head, and a first dorsal fin that originates closer to the pelvic free rear-tips than to the caudal fin origin. These are warm-temperate to tropical rays found on continental shelves and the uppermost continental slopes. They are mostly taken by recreational anglers in California waters but are of commercial importance in Mexican waters.

1a   Disc length greater than width; dorsal surface usually plain, occasionally with dark spots, but without prominent dark bars on back . . . .
. . . . Shovelnose Guitarfish (*Rhinobatos productus*)
1b   Disc length about equal to width; dorsal surface with several prominent dark bars on back . . . .
. . . . Banded Guitarfish (*Zapteryx exasperata*)

SHOVELNOSE GUITARFISH FOLLOWS ➤

SHOVELNOSE GUITARFISH   *Rhinobatos productus*

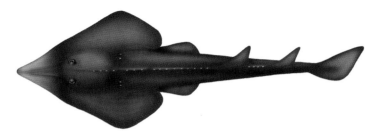

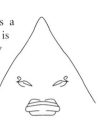

DESCRIPTION: The Shovelnose Guitarfish is a guitar-shaped ray with a broad disc that is greater in length than width; a relatively smooth dorsal surface except for a single row of thorns around the eyes and extending along the back and tail; a long, pointed snout with a rounded tip; small, rounded, pebblelike teeth; a first dorsal fin that originates closer to the pelvic fin base than to the caudal fin origin; a thick tail; and a moderately large caudal fin without a distinct lower lobe. The color ranges from an olive to sandy brown above, but without prominent dark bars across the back, and white below. Tooth counts: 102–112/ 98–117. Vertebral counts: 171–179. Spiral valve counts: 8–10.

HABITAT AND RANGE: Shovelnose Guitarfishes are a shallow-water species commonly found at a depth of 1 to 13 m, although they may be found at depths down to 91 m. They usually lie partially buried on sandy or mud bottoms but occasionally are observed in sea grass beds. These rays will at times congregate in large numbers in shallow bays and estuaries.

The Shovelnose Guitarfish is endemic to the eastern Pacific, ranging from San Francisco southward at least to the Gulf of California, and possibly to Mazatlan, Mexico.

NATURAL HISTORY: Viviparous, without a yolk sac placenta, with litters ranging from 6 to 28, although 9 to 11 is the average. The number of embryos per litter increases as the female grows in size. The size at maturity varies regionally among subpopulations along the coast. In southern California males mature at 90 to 100 cm and grow to a maximum length of 119 cm. Southern Baja subpopulations of males mature at a slightly smaller size, usually 80 cm. Females mature at 99 cm and grow to at least 145 cm, but are reputed to reach a maximum length of 170 cm. The size at birth is 15 to 23 cm.

Several shallow bays and estuaries in southern California and Baja are important areas for mating and pupping, and they also serve as a nursery ground for newborns. Females enter these bays during spring to complete gestation with birth occurring shortly thereafter in early summer. During the summer pupping season females may outnumber males by as much as 53 to 1 in these bays and estuaries. However, by midsummer, after the pupping season is nearly complete, males begin migrating into these same areas to mate and the sex ratio becomes about equal. After mating, adult rays of both sexes leave the area and are usually absent during fall and winter months. Gestation lasts about 12 months. Interestingly, the exceptionally warm waters that occur during El Niño years induce these rays to begin their reproductive cycle as much as a month or more earlier.

Males mature in eight years and live at least 11 years. Females mature in seven years and live at least 16 years.

Shovelnose Guitarfish feed on a variety of benthic invertebrates including polychaete worms, clams, amphipods, crabs, and shrimps. They crush clam shells with their jaws, spit out the prey item, and then consume the soft fleshy portions. Adults feed on bony fishes, which form a minor portion of their diet.

HUMAN INTERACTIONS: Shovelnose Guitarfish are taken incidentally in commercial and recreational fisheries, mostly in southern California. In Baja the Shovelnose Guitarfish is fished commercially. They are good eating and are readily sold in markets as food for human consumption. Coastal Native Americans in southern California consumed Shovelnose Guitarfish as a regular part of their diet.

Although considered harmless to humans there is one record of a scuba diver being bitten by an amorous male guitarfish that was following a female guitarfish in La Jolla Cove.

NOMENCLATURE: *Rhinobatos productus* (Girard, 1854). The genus name comes from the Greek *rhine,* meaning shark, and the Latin *batis,* meaning ray, in reference to its body form being intermediate between that of a shark and a ray. The species name comes from the Latin, meaning produced, in reference to its pointed snout. The common name refers its guitar-shaped body and broad, shovel-shaped disc. Locally it has also been referred to as the Guitarfish, Pointed-nose Guitarfish, or Shovel-nose Shark.

REFERENCES: Salazar-Hermosa and Villavicencio-Garayzar (1999); Talent (1982, 1985); Timmons and Bray (1997); Villavicencio-Garayzar (1993a); Zorzi and Martin (1995).

BANDED GUITARFISH   *Zapteryx exasperata*

DESCRIPTION: The Banded Guitarfish is a guitarfish-shaped ray with a broad disc that is about as wide as it is long; a dorsal surface covered with numerous, small to large, scattered, stellate prickles; a single median row of enlarged thorns running along the mid-back; a broad, short snout; small, blunt, pebblelike teeth; a first dorsal fin that originates closer to the pelvic fins bases than to the caudal fin origin; a thick tail; and a moderately large, rounded caudal fin without a distinct lower lobe. The dorsal surface is a sandy brown to dark gray, with several prominent black bars, and lighter below with dark spots on the posterior edge of the pectoral fins. Tooth counts: 60–75/60–75. Vertebral counts: 149–150. Spiral valve counts: 8–10.

HABITAT AND RANGE: This warm-temperate to tropical guitarfish is usually found on rocky reefs from the intertidal zone to a depth of 69 m.

The Banded Guitarfish is found from southern California to at least Mazatlán, Mexico and possibly Peru. Whether the Banded Guitarfish is actually found in Central and South American waters is uncertain. A second species, the Spotted Guitarfish (*Z. xyster*), is very common from Mexico to Peru and is often misidentified as the Banded Guitarfish.

NATURAL HISTORY: Viviparous, without a yolk sac placenta, with litters of 4 to 11 pups. Males mature at 64 to 70 cm and grow to a maximum size of 83 cm. Females mature at 57 to 77 cm and reach a maximum size of 97 cm. The size at birth is between 15 and 22 cm. Mating takes place around March and birth occurs three to four months later, usually in July, in the Pacific waters of Baja California. Males and females appear to be highly segregated, with females tending to congregate in the shallower waters of bays and lagoons. Mixed schools of adults are found only in March and April when mating occurs. Several lagoons along the Pacific coast of Baja serve as important nursery grounds.

Banded Guitarfish feed on crustaceans, including shrimps and crabs, and other benthic invertebrates. During the day they are usually found resting in caves or under ledges, but at night they become quite active as they forage for food on and around rocky reefs.

HUMAN INTERACTIONS: Banded Guitarfish are occasionally taken as a by-catch in commercial and recreational fisheries but are of no commercial importance in California waters. They are a commercially important species in Mexican waters, particularly in the Gulf of California. These are relatively docile, harmless rays that are easily approached by divers.

NOMENCLATURE: *Zapteryx exasperata* (Jordan and Gilbert, 1880). The genus name comes from the Greek *za*, meaning intensive, and *pteryx*, meaning fin, in reference to the vertical fins being larger than those of skates. The species name comes from the Latin, meaning made rough, in reference to the numerous stellate prickles on its back. Its common name refers to the prominent dark bars on its back and its overall guitar-shaped body. Locally it has also been referred to as the Mottled Guitarfish, prickly skate, or striped guitarfish.

When first described by Jordan and Gilbert (1880b) the Banded Guitarfish was placed in the genus *Platyrhina*. However, Jordan and Gilbert later erected the genus *Zapteryx* and moved this species into it. The Spotted Guitarfish (*Z. xyster*), although distinct, is often confused with the Banded Guitarfish. The Spotted Guitarfish can be distinguished from the Banded Guitarfish by the presence of several very prominent yellow spots on its back.

REFERENCES: Villavicencio-Garayzar (1995a).

Thornback Rays (Platyrhinidae)

The thornback rays are represented by two genera and at least three species. Only one species inhabits California waters. The head, pectoral fins, and trunk of members of this family form a heart-shaped disc. They have from one to three prominent rows of small to large thorns on their back and predorsal tail. The first dorsal fin midbase is closer to the caudal fin origin than to the pelvic fin bases. These are warm-temperate rays found on soft sediment bottoms of continental shelves. Very little is known about their general biology.

THORNBACK RAY   *Platyrhinoidis triseriata*

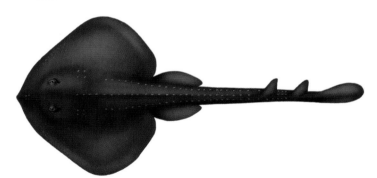

DESCRIPTION: The head, pectoral fins, and trunk of these rays form a heart-shaped disc; they have a short, broadly rounded snout; between one and three large, hooked-shaped thorns extending across the midback to the tail; small, pebblelike teeth; a first dorsal fin originating closer to a caudal origin than to the pelvic fin free rear-tips; a thin tail; and a caudal fin without a distinct lower lobe. The dorsal surface is a light olive to gray brown without any distinct markings;

the lower surface is a cream white. Tooth counts: 68–82/64–78. Vertebral counts: 130–133. Spiral valve counts: 6–8.

HABITAT AND RANGE: The Thornback Ray is a common inshore species usually found in water less than 6 m deep, although they have been taken down to a depth of 137 m. They are usually found on the mud and sandy bottoms of bays, sloughs, and coastal beaches and around kelp forests. At times of the year they will concentrate in large numbers in coastal bays and sloughs.

Thornback Rays range from Tomales Bay to the Gulf of California, although they are uncommon north of Monterey Bay. They are endemic to the eastern Pacific.

NATURAL HISTORY: Viviparous, without a yolk sac placenta, with litters between 1 and 15. Males mature at 37 cm and females at 48 cm. The maximum recorded size is 91 cm. Size at birth is about 11 cm. Mating takes place in late summer and birth occurs the following summer, usually in August.

Thornback Rays feed on polychaete worms, crabs, shrimps, squids, and small bony fishes including anchovies, gobies, sardines, sculpin and surfperch. Sharks and Northern Elephant Seals are known predators.

HUMAN INTERACTIONS: Thornback Rays are occasionally caught in commercial and recreational fisheries but are of no commercial importance.

These are fairly docile rays and are easily approached by divers.

NOMENCLATURE: *Platyrhinoidis triseriata* (Jordan and Gilbert, 1880). The genus name comes from the Greek *platys,* meaning broad, and *rhinos,* meaning snout. The species name comes from the Latin *tres,* meaning three, and *series,* meaning row, in reference to the three rows of hooklike thorns extending along its back and tail. Its common name refers to the numerous sharp, enlarged spines on its back. Locally it has also been referred to as the Round Skate, Thornback Guitarfish, or Thornback.

The Thornback Ray was originally placed in the genus *Platyrhina* by Jordan and Gilbert (1880c) but was subsequently placed in *Platyrhinoidis* by Garman (1881), who erected the new genus.

REFERENCES: Feder et al. (1974); Plant (1989).

Torpedo Rays (Torpedinidae)

The torpedo rays consist of a single genus with at least 27 known species. A single species occurs in California waters. These rays are characterized by a rounded disc; short truncated snout; two moderately large dorsal fins; a short, stout tail; and a large caudal fin. These are temperate to tropical rays found on continental shelves, usually in shallow water, but sometimes at depths down to 550 m. They feed mostly on fishes. There is a small commercial fishery for these rays, which are used in biological and medical studies.

PACIFIC TORPEDO RAY   *Torpedo californica*

DESCRIPTION: The Pacific Torpedo Ray is a soft, flabby-bodied ray, with an oval disc; a smooth dorsal surface, with a visible kidney-shaped electric organ; small teeth

with a single smooth-edged cusp; a first dorsal fin that is nearly twice the size of the second; a short, stocky tail; and a large caudal fin. Electric rays are a uniform dark gray to bluish or brown above, occasionally with small black spots, and lighter below. Tooth counts: 25–28/19–26. Vertebral counts: 98–105. Spiral valve count: 10–12.

HABITAT AND RANGE: The Pacific Torpedo Ray is found on sandy bottoms, around rocky reefs, and near kelp beds. They are commonly found at a depth of 3 to 30 m in California waters but along the Baja coast are most commonly observed from 100 to 200 m deep. A Pacific Torpedo Ray was once videotaped 17 km west of Point Pinos, Monterey County, swimming at a depth of 10 m over water 3,000 m deep. Similar observations along the coast suggest that these torpedo rays periodically will move offshore into an epipelagic habitat. They prefer water temperatures between 50 and 55 degrees F.

The Pacific Torpedo Ray is endemic to the eastern North Pacific ranging from Magdalena Bay, Baja California, to northern British Columbia, possibly with one or more discrete populations north of Point Conception.

NATURAL HISTORY: Viviparous, without a yolk sac placenta, with litters of 17 to 20. Males mature at about 65 cm and reach a maximum length of 92 cm. Females mature at about 73 cm and grow to a maximum size of 137 cm. Size at birth is 18 to 23 cm. There is no defined birthing season. Their reproductive cycle is believed to be annual in males and biannual in females.

Males mature in about seven years and females in about nine years. They live to at least 16 years and possibly up to 24 years. Pacific Torpedo Rays nearly double their size in the first year of life, growing about 25 cm during this time.

Pacific Torpedo Rays feed mainly on fishes, including anchovies, hake, herring, mackerel, flatfishes, and Kelp Bass, but also will feed on invertebrates including cephalopods. Pacific electric rays forage for food using two main strategies, depending on whether they are hunting at night or during the day. At night or in turbid conditions they actively forage in the water column searching for fish and other potential prey items. Because their vision is somewhat limited at night or in turbid water, they rely on electrical cues or stimuli given off by prospective prey items. This gives these rays a hunting advantage over their prey. During the day Pacific Torpedo Rays will quietly rest on the bottom, partially buried in the sand, lying in wait to ambush an unsuspecting prey item that swims within striking distance. Prey captured during the day happens more by chance, as these rays are resting, as opposed to at night when they are actively foraging.

HUMAN INTERACTIONS: Pacific Torpedo Rays are taken in a small commercial fishery in southern California for the purposes of biological and medical research. They are taken as a by-catch in commercial and recreational fisheries but are of no commercial value.

Caution should be exercised by anglers and divers who may encounter these rays as they can discharge a powerful electric shock of 45 volts or more, strong enough to knock down a grown adult. At night, torpedo rays are extremely active and will swim directly at divers if confronted or harassed. There are no known confirmed fatalities by these rays in California waters, although there are several suspicious, unexplained, fatal scuba diving accidents that may have involved these rays.

NOMENCLATURE: *Torpedo californica* (Ayres, 1855). The genus name comes from the Latin *torpidus,* meaning numbness, in reference to the numbing effect from its electric organs. The species name comes from where it was first described. Locally it is also known as the California Torpedo Ray, Torpedo Ray, or Electric Ray.

The Pacific Torpedo Ray was first described as *T. californica* by Ayres (1855b), but was subsequently placed in the genus *Tetronarce* by Gill (1861), who stated that the genus had been preoccupied by another species, thus invalidating it. Subsequent taxonomic research validated *Torpedo* as the proper genus for this species.

REFERENCES: Bray and Hixon (1978); Feder et al. (1974); Lowe et al. (1994); Neer and Cailliet (2001).

Softnose Skates (Arhynchobatidae)

The softnose skates are composed of 11 genera and 81 to 85 species. A single genus and six or seven species, one of which may be undescribed, are found in California waters. These skates are distinguished by their short, flexible snout. Most of the members of this family are found in deep water, usually on the outer continental shelf and upper slopes. All of the California species are deep living, with some known from only a few individual specimens. Most are taken in deepwater fisheries as a by-catch, but little else is known about their occurrence in California waters. Little is known about the taxonomic placement and distribution for many of these skates as they exhibit a high degree of morphological variation throughout their range and, in fact, may actually represent several species.

1a   Scapular thorns present 2
1b   Scapular thorns absent 3

      2a   Dorsal surface with a continuous row of median thorns; ventral surface is white except for snout, posterior disc margins, pelvic fins, anal area, and underside of tail, which are a prominent dark brown or gray . . . . Aleutian Skate (*Bathyraja aleutica*)

      2b   Dorsal surface with a noncontinuous row of median thorns; ventral surface, including tail, is white without

---

DEEPSEA SKATE   *Bathyraja abyssicola*

DESCRIPTION: The Deepsea Skate is a softnose skate with a moderately triangular anterior disc margin; a broadly rounded posterior disc margin; a disc width slightly greater than its length; rounded pectoral fin apices; one to five nuchal thorns sepa-

rated from a row of 21 to 31 continuous median tail thorns; a moderately long, narrow, tapering tail; two relatively large, close-set dorsal fins, usually with an interdorsal thorn present; and a small, low-set caudal fin that is distinctly separated from the second dorsal fin. The dorsal surface is a grayish purple to dark chocolate brown or black, occasionally with scattered small darkened spots; the anterior tip of the pelvic fins is whitish, with the ventral surface slightly darker, except for a whitish area around the mouth. Large males have irregular whitish blotches and numerous dark spots; in females, the whitish blotches are reduced or absent. Juveniles tend to be uniform in color. Tooth counts: 30–36/30–36. Predorsal vertebral counts: 70–78. Spiral valve counts: 8–11.

HABITAT AND RANGE: The Deepsea Skate, as the name implies, occurs in deep water, usually on the continental slope, at a depth of 362 to 2,906 m.

They range from off northern Baja, around Coronado Island and Cortes Bank, north to the Bering Sea, and as far west as Japan.

NATURAL HISTORY: Oviparous; egg cases from this species have never been observed. Males mature at 110 to 120 cm and grow to at least 135 cm. Females grow to at least 157 cm, but their size at maturity has not been established. Size at birth is uncertain, although the smallest free-swimming specimens measured 34 to 36 cm.

The Deepsea Skate is a voracious predator on benthic invertebrates including annelid worms, cephalopods, tanner crabs, shrimps, and bony fishes. Smaller individuals (less than 1 m) tend to consume benthic invertebrates in relatively higher proportion than adults, which prefer bony fishes.

HUMAN INTERACTIONS: The Deepsea Skate is a fairly common below 1,000 m. It is a very rare species known from only a few dozen specimens. It is taken incidentally as a by-catch in deepwater trawls and traps.

NOMENCLATURE: *Bathyraja abyssicola* (Gilbert, 1896). The generic name *bathy* comes from the Greek *bathos,* meaning deep, and *raja,* meaning skate. The species name *abyssos* is from the Greek, meaning bottomless, and *cola,* meaning living at depths. The common name is in reference to the deepsea habitat of this skate.

The genus *Bathyraja* was originally described as a subgenus by Ishiyama (1958) but was later elevated to full generic status by Ishiyama and Hubbs (1968). The Deepsea Skate, like other members of this genus, was originally placed in the genus *Raja* when first described.

REFERENCES: Zorzi and Anderson (1988); Zorzi and Martin (1994).

ALEUTIAN SKATE   *Bathyraja aleutica*

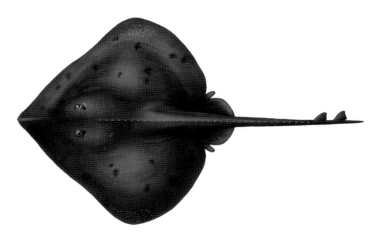

DESCRIPTION: The Aleutian Skate is a soft-nose skate with a disc width slightly longer than wide; a long snout, nearly one-third the disc length, and broadly triangular; a deeply concave space between the eyes; pectoral fin apices broadly rounded; two similar-sized dorsal fins, occasionally with one or two interdorsal thorns present; a tail less than half its total length; a dorsal surface uniformly covered with fine minute prickles; a single, continuous row of 34 to 40 prominent median thorns, uniform in size; scapular thorns numbering one or two on each shoulder; and a smooth ventral surface. It is dark brown to olive or gray above with faint dark spots on its pectorals; the ventral surface is white except for the snout, posterior disc margins, pelvic fins, anal area, and underside of the tail, which are dark brown or gray. Tooth counts: 31–39/31–43. Predorsal vertebral counts: 67–90. Spiral valve counts: 8–11.

HABITAT AND RANGE: A deepwater skate, the Aleutian is found along the outer continental shelf and upper slopes on muddy bottoms at a depth of 91 to 700 m. These skates have been caught in water at temperatures as low as 36 degrees F.

Aleutian Skates are endemic to the North Pacific, ranging from Cape Mendocino, northern California, to the Bering Sea and northern Japan.

NATURAL HISTORY: Oviparous; the egg cases are relatively large and are covered on the surface by numerous long prickles that are rough to the touch, with long horns at each of the four corners. Males are mature at 113 cm and reach a maximum length of at least 150 cm. Females mature at 125 cm and grow to a maximum length of about 154 cm. Size at birth is 12 to 15 cm. Nursery grounds for this species along the California coast, like other local skate species, are unknown. However, there does appear to be an important Aleutian Skate nursery ground in a region along the southeastern continental slope of the Bering Sea at a depth of 250 to 500 m. Egg cases are deposited continuously between June and November, but the incubation time for the embryos to develop is unknown.

Aleutian Skates feed on a wide variety of benthic invertebrates, including annelid worms, crustaceans such as crabs and shrimps, and bony fishes including pollock and Sockeye Salmon.

HUMAN INTERACTIONS: Aleutian Skates are occasionally taken as a by-catch in northern California but are of no commercial importance.

NOMENCLATURE: *Bathyraja aleutica* (Gilbert, 1896). The common name and species name refer to the Aleutian Islands, where the holotype was collected.

REFERENCES: Hoff (2002); Teshima and Tomonaga (1986).

SANDPAPER SKATE
Bathyraja kincaidii

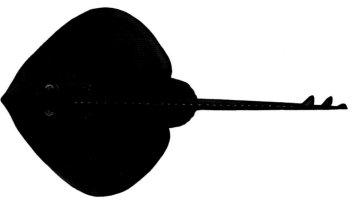

DESCRIPTION: The Sandpaper Skate is a softnose skate with a disc width greater than its length; a short snout, broadly rounded; a flat or slightly concave

space between the eyes; rounded pectoral fin apices; a dorsal surface uniformly covered with small prickles, sandpaper-like to the touch; a smooth ventral surface; a single, noncontinuous row of 27 to 33 median thorns; scapular thorns numbering one or two on each shoulder; two similar-sized dorsal fins, occasionally with an interdorsal thorn present; a tail as long as its disc; and a small caudal fin distinctly separated from the second dorsal fin. Adults are dark brown to gray above and white below, including the tail. Juveniles have numerous small dark spots above; the tail usually has white spots on either side. Tooth counts: 20–33/24–31. Predorsal vertebral counts: 64–73. Spiral valve counts: 9–11.

HABITAT AND RANGE: A deepwater skate, the Sandpaper is found most commonly at a depth of 200 to 500 m. These skates are usually found in deeper water in the southern portion of their range.

Sandpaper Skates are endemic to the North Pacific, ranging from northern Baja California to British Columbia and possibly to the Gulf of Alaska. It has been observed in the Bering Sea, although this most likely is a different species.

NATURAL HISTORY: Oviparous; the egg cases are moderately rough to the touch, with prominent lateral keels and long inwardly bent anterior horns. The Sandpaper Skate produces the smallest egg case of any California skate species. Males are mature by 48 cm and reach a maximum length of 53 cm. Females mature at 46 to 50 cm and grow to a maximum length of about 56 cm. Size at birth is 12 to 16 cm.

Sandpaper Skates feed on a wide variety of benthic invertebrates, including polychaete worms, amphipods, and small crustaceans such as crabs and shrimps.

HUMAN INTERACTIONS: Although occasionally taken as a by-catch, the Sandpaper Skate is of no commercial importance. They have been maintained in captivity at public aquariums.

NOMENCLATURE: *Bathyraja kincaidii* (Garman, 1908). The species name is in honor of Dr. Trevor Kincaid, University of Washington, who collected the holotype specimen from Friday Harbor, Washington. The Sandpaper Skate is so named because of the coarse, uniformly covered denticles on its dorsal surface. It has also been referred to as the Black Skate.

The Sandpaper Skate has been synonymized by some authors as *Rhinoraja interrupta,* a rare skate known from the Bering Sea and Gulf of Alaska, although current taxonomic evidence suggests that these two species are distinct. The Sandpaper Skate appears to have at least one or more morphological variants, each possibly representing a distinct subspecies throughout its range. A detailed systematic review of this species is required to determine its status.

REFERENCES: Zorzi and Martin (1994).

FINE-SPINED SKATE   *Bathyraja microtrachys*

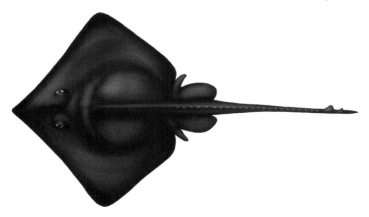

DESCRIPTION: The Fine-spined Skate is a soft-nose skate with a disc that is slightly wider than long; broadly rounded pectoral fin apices; a disc dorsal surface evenly covered with numerous, small prickles; no enlarged thorns; a tail with a single median row of 22 to 23 thorns; a smooth ventral surface; a tail slightly longer than its disc; two similar-sized dorsal fins with a minute interdorsal thorn; a distinct space between the dorsal fins; a second dorsal fin free rear-tip that slightly overlaps the caudal fin; and a short tapering caudal fin with a filamentous fold on its upper surface. In preservative the color is a uniform brown above, being slightly darker at the disc margins. The ventral surface is white except for the "wings" and the pelvic region, which are brown. Tooth counts: 29–32/25–26. Predorsal vertebral counts: 69–70. Spiral valve count: 9.

HABITAT AND RANGE: The Fine-spined is a rare Skate found only in very deep water between 1,995 and 2,900 m. It appears to be fairly common below 2,000 m, but confusion with other deepwater skates such as the Deepsea Skate and Roughtail Skate has precluded a better understanding of this species. It appears to assume the ecological niche of the Roughtail Skate at the great depths at which it lives.

This little-known skate is endemic to the eastern North Pacific, found sporadically from about 300 miles southwest of San Diego to off Washington state.

NATURAL HISTORY: Oviparous; egg cases are unknown for this species. Females mature at 60 to 70 cm. The largest recorded male is an immature specimen measuring 35 cm. Size at birth is about 17 cm.

Little is known about their diet other than they feed on deep-water shrimps.

HUMAN INTERACTIONS: They are occasionally taken as a by-catch in deepwater fisheries.

NOMENCLATURE: *Bathyraja microtrachys* (Osburn and Nichols, 1916). The species name is derived from the Latin *micro,* meaning small, and *trachys,* meaning spine, in reference to the uniformly fine prickles covering the dorsal surface of its body. The common name also refers to these fine prickles.

The Fine-spined Skate is often confused with the Roughtail Skate and, until recently, was considered a junior synonym of that species. However, the white on its ventral surface extending from the snout to the pelvic region flanked by its brownish "wings" readily distinguishes it from the Roughtail Skate and all other skates found within its known range.

REFERENCES: Osburn and Nichols (1916); Townsend and Nichols (1925).

PACIFIC WHITE SKATE · *Bathyraja spinosissima*

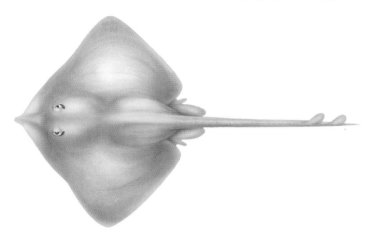

DESCRIPTION: The Pacific White Skate is a softnose skate with a disc that is slightly wider than long; broadly rounded pectoral fin apices; a disc surface evenly covered with numerous,

small prickles above and below, giving it a shagreenlike texture to the touch; no enlarged thorns except for alar spines of adult males and no median thorns on the back; a tail slightly longer than its disc, with a single median row of 23 to 29 thorns; two similar-sized dorsal fins without an interdorsal thorn; a space between the second dorsal and caudal fins; and a long, tapering caudal fin with a filamentous fold on its upper surface. It is a uniform pale to salty gray above and below, with dusky outer disc margins. Tooth counts: 34/23. Predorsal vertebral counts: 82. Spiral valve count: 12–13.

HABITAT AND RANGE: Found at a depth of 800 to 2,938 m, this is one of the deepest living of all skate species.

The Pacific White Skate ranges from the Galapagos Islands to off Waldport, Oregon. The only confirmed California record is of an egg case and embryo taken in a trawl off the Farallon Islands. Elsewhere it was reported in the Sea of Okhotsk off northern Japan, but this may be a different species.

NATURAL HISTORY: Oviparous; the egg cases are longitudinally striated and olive green in color. Size at birth is about 25 cm. Size at maturity is unknown. Maximum length is about 1.5 m.

Little is known about their diet other than that they feed on benthic fishes.

HUMAN INTERACTIONS: They are occasionally taken as a by-catch in deepwater fisheries.

NOMENCLATURE: *Bathyraja spinosissima* (Beebe and Tee-Van, 1941). The species name is derived from the Latin *spinosus,* meaning thorny, in reference to the prickles uniformly covering most of its body. The common name refers to its uniformly light coloration.

REFERENCES: Zorzi and Martin (1994).

ROUGHTAIL SKATE   *Bathyraja trachura*

DESCRIPTION: The Roughtail Skate is a soft-nose skate with a disc width greater than disc length; broadly rounded pectoral fin apices; a dorsal surface covered with thorns and smaller prickles; a smooth ventral surface; nuchal thorns usu- ally absent, but if present less than three and weakly developed; a short, stout tail, its length less than the disc length in adults; a median row of 15 to 34 tail thorns; and two similar-sized dorsal fins with no interdorsal thorn. It is uniformly plum brown to black and occasionally lighter below. Tooth counts: 26–35/ 22–35. Predorsal vertebral counts: 62–66. Spiral valve count: 6–9.

HABITAT AND RANGE: A deepwater skate, it is found at depths of 400 to 2,550 m.

The Roughtail Skate ranges from northern Baja to the western Bering Sea and the Sea of Okhotsk.

NATURAL HISTORY: Oviparous; the egg cases are smooth, plush-like to the touch, with long, slender, pointed horns curving inward to slightly overlap at their tips. Males mature at about 75 cm with a maximum length of at least 83 cm. Females mature at 74 to 84 cm with a maximum length of at least 89 cm. Size at birth is 9 to 16 cm.

Roughtail Skates feed mainly on benthic invertebrates including annelid worms, shrimps, and crabs. Adults prefer small fishes such as rattails and flatfishes. The egg cases of these skates are preyed upon by molluscs that bore holes in them to feed on the protein-rich yolk sac.

HUMAN INTERACTIONS: They are occasionally taken as a by-catch but are of no commercial value.

NOMENCLATURE: *Bathyraja trachura* (Gilbert, 1892). The species name comes from the Greek *trachys*, meaning rough, in reference to its thorny dorsal surface and tail. The common name also reflects this characteristic. The Roughtail Skate is also referred to as the Black Skate.

REFERENCES: Zorzi and Martin (1994).

Hardnose Skates (Rajidae)

The hardnose skates are composed of 15 genera and 133 to 138 species, a number likely to increase as new species are described. Two genera and five species occur in California waters. These skates are distinguished by their stout, stiff, elongated snout. They generally range from close inshore, inside bays, to the outer continental shelf and upper slopes. Hardnose skates are generally found in deeper water in the southern portion of their range. The group includes some of the most commercially important skates.

1a   Scapular thorns present ....
.... Broad Skate (*Amblyraja badia*)

1b Scapular thorns absent 2
    2a Pelvic fins shallowly notched . . . .
    . . . . Big Skate (*Raja binoculata*)
    2b Pelvic fins deeply notched 3
3a Snout very short, bluntly pointed. . . . .
    . . . . Pacific Starry Skate (*Raja stellulata*)
3b Snout long, acutely pointed. 4
    4a Snout extremely long, tapering to an acute point; ventral surface mottled light to dark gray . . . .
    . . . . Longnose Skate (*Raja rhina*)
    4b Snout moderately long, acutely pointed; ventral surface white to pale tan. . . . . . California Skate (*Raja inornata*)

BROAD SKATE　　　　　　　　　　　　　　　*Amblyraja badia*

DESCRIPTION: The Broad Skate is a hardnose skate with a disc width greater than its length; sharply rounded pectoral fin apices; a short, blunt snout with several enlarged thornlets on its tip; a dorsal surface with densely covered prickles; a smooth ventral surface; two or three pairs of scapular thorns usually arranged in a triangle on each shoulder; 24 to 29 median thorns in a single continuous row; a tail with a single row of smaller, lateral thornlets flanking the median thorns; a relatively short, narrow tail tapering to the tip; dorsal fins similar in size, without an interdorsal thorn; and a very small caudal fin close-set behind the second dorsal fin. The dorsal surface is chocolate-brown to gray-brown with scattered darker spots. The ventral surface is the same as the dorsal surface except for the pelvic fin lobes and tail, which are noticeably darker. There are whitish areas on the snout, upper abdomen, nostrils, mouth, gill slits, and anal opening. Tooth counts: 37–45/38–42. Predorsal vertebral counts: 52–57. Spiral valve counts: 9.

HABITAT AND RANGE: This rare skate is found in very deep water from 846 to 2,324 m.

The Broad Skate is sporadically distributed in the eastern Pacific from Panama to the Navarin Canyon, Bering Sea, and possibly from off northern Japan.

NATURAL HISTORY: Oviparous; the egg cases are unknown for this species. Males mature at 86 to 93 cm and grow to at least 95 cm. Females grow to at least 99 cm, but size at maturity is unknown. The smallest known free-swimming specimen measured 23 cm.

Broad Skates are known to feed on cephalopods, crustaceans, and small bony fishes such as rattails.

HUMAN INTERACTIONS: They are occasionally taken as a by-catch in deepwater trawls and traps.

NOMENCLATURE: *Amblyraja badia* (Garman, 1899). The generic name comes from the Greek *amblys,* meaning blunt, and the Latin *raja,* meaning skate. The species name comes from the Latin *batius,* meaning brown, in reference to its coloration. The common name comes from its distinctly broad disc.

The Broad Skate was previously in the genus *Raja,* subgenus *Amblyraja,* until the latter was elevated to full generic status based on a systematic revision of this skate group.

REFERENCES: Zorzi and Anderson (1988); Zorzi and Martin (1994).

BIG SKATE   *Raja binoculata*

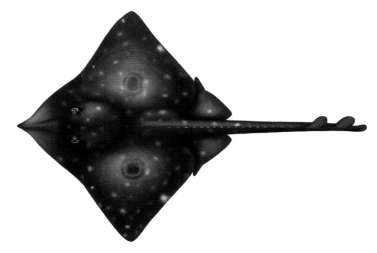

DESCRIPTION: The Big Skate is a hardnose skate with a disc width slightly greater than its length; a long and moderately pointed snout; a dorsal surface covered with small prickles, smooth in juveniles; a ventral surface with prickles on the

snout, between the gills, and on the abdominal region; a midback with zero to two thorns; a single row of 12 to 55 (usually 13 to 17) median thorns extending along the tail; pelvic fins weakly notched; a space between the dorsal fins with an interdorsal thorn; and a caudal fin that is reduced to little more than a finfold. The dorsal surface is brown to reddish brown, olive-brown, or gray, with darker mottling and rosettes of white spots and two prominent eyespots. The ventral surface is white, with occasional dark spots or blotches. Tooth counts: 24–48/22–45. Predorsal vertebral counts: 96. Spiral valve counts: 11–15.

HABITAT AND RANGE: The Big Skate is a coastal skate species commonly found on the continental shelf from inside shallow bays to a depth of 800 m, although it is usually found at more moderate depths.

Big Skates range from Cedros Island, central Baja California, to the eastern Bering Sea, occurring in progressively shallower water in the northern portion of their range.

NATURAL HISTORY: Oviparous; these are the largest egg cases of any California skate. In addition to their size these egg cases can be recognized by their lateral edges, which are parallel posteriorly but become somewhat concave near the middle of the case; four nearly equal length, short, blunt, broad horns with the posterior pair being slightly longer; and a dorsal surface that is highly arched and a ventral surface that is relatively flat. Each egg case may contain one to seven embryos, with three or four being most common. This is the only skate egg case that routinely contains more than a single embryo. Females mature at about 1.3 m and males at 1.0 to 1.1 m. Maximum size is reported to be 2.4 m, but individuals over 1.8 m are uncommon. Size at birth is 18 to 23 cm. There does not appear to be a defined breeding season as egg cases are deposited year-round. Males mature in about 10 to 11 years and females in about 12 years.

Big Skates feed on polychaete worms, molluscs, crustaceans, and small benthic fishes. Juveniles tend to consume polychaete worms and molluscs in slightly higher proportion than adults. Sevengill Sharks and Northern Elephant Seals are known predators on these skates.

HUMAN INTERACTIONS: Usually taken as a by-catch, this is one of the most commercially important skates landed in California waters. The meat from the "wings" is marketed for human consumption. They are frequently taken in San Francisco Bay by recreational anglers, usually in slightly higher numbers during winter and spring than at other times of the year.

NOMENCLATURE: *Raja binoculata* (Girard, 1854). The generic name *Raja* comes from the Latin meaning a skate. The species name comes from the Latin *bi,* meaning two, and *oculatus,* meaning eyed, in reference to the prominent eyespot at the base of each pectoral fin. It is the largest of California's known skate species, hence its common name refers to its large size.

A junior synonym, *R. cooperi,* was described by Girard (1858) based on the notes and sketches made by James G. Cooper of a large, decaying Big Skate that had washed ashore near the entrance of Shoalwater Bay, Washington.

REFERENCES: Zeiner and Wolf (1993).

CALIFORNIA SKATE   *Raja inornata*

DESCRIPTION: The California Skate is a hardnose skate with a disc width slightly wider than its length; moderately rounded apices; a moderately long and acutely pointed snout; a dorsal surface with small, scat- tered prickles; a smooth ventral surface; zero to seven nuchal thorns; a tail with a median row of 10 to 66 thorns flanked laterally on either side by a row of smaller thornlets; tail length shorter than disc length; an interdorsal thorn between two similar-sized dorsal fins; and a very small caudal fin. It is olive brown above with occasional darker mottling and eyespots; the ventral surface is lighter. Tooth counts: 36–45/34–42. Predorsal vertebral counts: 42–46. Spiral valve counts: 7–9.

HABITAT AND RANGE: This is a common nearshore species found on soft bottoms at a depth of 17 to 671 m.

The California Skate is found from the Straits of Juan de Fuca, Washington state, to central Baja California, with a disjunct population residing in the Gulf of California.

NATURAL HISTORY: Oviparous, with egg cases laid year-round. Males mature at about 47 cm and grow to about 60 cm. Females mature at about 52 cm, with a maximum length of 76 cm. The size at birth is 15 to 23 cm.

California Skates feed on small benthic invertebrates including polychaete worms and shrimp.

HUMAN INTERACTIONS: This is a commercially important skate usually taken as a by-catch in other fisheries.

NOMENCLATURE: The species name comes from the Latin *inornatus,* meaning unadorned, in reference to the smoothness of its disc. The common name is in reference to the type locality.

Raja inornata inermis (Jordan and Gilbert, 1881) was based on a supposed variant of *R. inornata* caught off Santa Barbara, but with no specifics given as to why it was distinct. *Raja jordani* (Garman 1885) was based on an adult male *R. inornata* caught off San Francisco. Both species are junior synonyms of *R. inornata.*

REFERENCES: Martin and Zorzi (1993).

LONGNOSE SKATE   *Raja rhina*

DESCRIPTION: The Longnose Skate is a hardnose skate with a wide disc; sharply rounded pectoral fin apices; a sexually dimorphic disc; an extremely long snout tapering to an acute point; a dorsal surface with sharply pointed prickles, smooth in young juveniles; a ventral surface prickled in females and smooth in males; a midback with one to three median thorns; 11 to 43 tail thorns with lateral rows present in very large individuals; small teeth with a single cusp; deeply notched pelvic fins; a short tail whose length is less than that of the disc; two similar-sized dorsal fins with zero to three interdorsal thorns; and a caudal fin that is reduced to little more than a finfold. The dorsal surface is light to dark brown or gray, with or without numerous irregular dark spots and blotches increasing in size medially. The ventral surface is mottled light to dark gray. Tooth counts: 37–49/35–46. Predorsal vertebral counts: 47–53. Spiral valve counts: 9–12.

HABITAT AND RANGE: The Longnose is a common skate found from nearshore to a depth of 1,000 m. They prefer a mud-cobble bottom near boulders, rock ledges, and other areas with some type of vertical relief.

Longnose Skates range from the southeastern Bering Sea southward to Cedros Island, Baja California, and the Gulf of California.

NATURAL HISTORY: Oviparous; the egg case surface is rough with a loose covering of attachment fibers and short horns. Males reach sexual maturity at 62 to 74 cm and grow to at least 105 cm. Females mature at 70 to 100 cm and reach a maximum size of 137 cm. Size at birth is 12 to 17 cm.

Males and females grow rapidly the first three years, but growth slows thereafter. Males mature in 10 to 11 years and females in 10 to 12 years.

Longnose Skates feed mainly on benthic crustaceans and bony fishes. Those individuals over 60 cm feed more on bony fishes, and those under 60 cm feed more on crustaceans. Although this is unconfirmed they appear to capture prey species typically found on or near reefs with some vertical relief. They catch these prey species by lying in wait and ambushing those that stray off the reef.

Longnose Skates are preyed upon by sharks and Sperm Whales. The egg cases of these skates are preyed upon by molluscs, which bore holes in them to feed on the protein-rich yolk sac. Egg cases of this skate have also been found in the stomachs of Sperm Whales.

HUMAN INTERACTIONS: They are taken as a by-catch in trawl fisheries and on long-lines but are of little commercial value as the meat is of poor quality.

NOMENCLATURE: *Raja rhina* (Jordan and Gilbert, 1880). The species name comes from the Greek *rhino*, meaning snout, in

reference to its long snout. The common name is also in reference
to its elongated snout.

REFERENCES: Zeiner and Wolf (1993).

PACIFIC STARRY SKATE

Raja stellulata

DESCRIPTION: The Pacific Starry Skate is a
hardnose skate with a disc width broader
than its length; a very short, bluntly
pointed snout; a dorsal surface covered with numerous, small, star-shaped

prickles, except at the pectoral fin bases of males; a ventral surface
usually smooth except around the snout; thorns on the midback
continuous with tail thorns, numbering 20 to 90 or more, and increasing with skate size; teeth with three small blunt cusps; pelvic
fins deeply notched; a slender tail tapering posteriorly, with a
length slightly shorter than disc length; two similar-sized dorsal
fins with zero or one interdorsal thorn; and a caudal fin that is
reduced to little more than a small dorsal finfold. The dorsal surface is brown to gray-brown, with numerous dark spots, becoming lighter upon death. An eyespot is often at the base of each
pectoral fin, with a prominent white spot behind each. The ventral surface is whitish except for dark brown to gray margins.

Tooth counts: 32–37/ 27–40. Predorsal vertebral counts: 119. Spiral valve counts: 9–11.

HABITAT AND RANGE: This is a common nearshore skate found at moderate depths, usually less than 100 m, although it may be found at depths of 732 m. It is very abundant in Monterey Bay during winter and spring.

The Pacific Starry Skate ranges from northern Baja California to Eureka, California. Specimens from Alaska and the Bering Sea are most likely a different species, possibly the Alaska Skate (*Bathyraja parmifera*).

NATURAL HISTORY: Oviparous; egg cases are striated with long robust horns. Very little is known about their reproductive biology except that males mature at about 67 cm and females at about 68 cm; the maximum size for both sexes is 76 cm. Size at birth is 12 to 16 cm.

Starry Skates feed primarily on benthic shrimps, cephalopods, and bony fishes including small Lingcod and rockfishes.

HUMAN INTERACTIONS: They are occasionally taken as by-catch.

NOMENCLATURE: *Raja stellulata* (Jordan and Gilbert, 1880). The species name comes from the Latin *stella*, meaning a small star, in reference to the numerous star-shaped prickles covering its dorsal surface. This is also reflected in its common name.

Raja montereyensis (Gilbert 1915), a junior synonym of this species, was based on an immature female caught off Santa Cruz.

REFERENCES: Jordan and Gilbert (1880d).

Round Stingrays (Urolophidae)

The round stingrays are composed of four genera and 36 described species worldwide. In California waters the family is represented by a single species, although two other species, the Spotted-round Stingray (*Urobatis maculatus*) and the Lined-round Stingray (*Urotrygon rogersi*), known from central Baja, may be found during periods of extremely warm water. Round stingrays are primarily found in shallow tropical and warm-temperate waters usually less than 70 m deep, although some are found at depths below 100 m. These are small- to medium-sized stingrays mostly ranging from 20 to 50 cm in disc width. These rays are occasionally taken as a by-catch but are of little commercial value. Although not fatal, a sting from the spine of these rays can be quite painful.

ROUND STINGRAY   *Urobatis halleri*

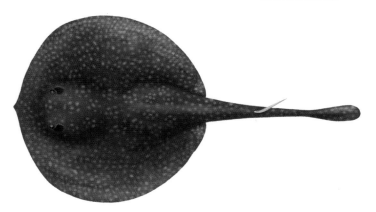

DESCRIPTION: The Round Stingray is a small stingray with a nearly round disc; small, sexually dimorphic, diamond-shaped teeth, with the central teeth of  males being erect, sharply pointed, and curved inward; a long, thick, serrated stinging spine; and a short, stout tail that is less than the disc length of these rays. The dorsal surface is usually brown or gray-ish brown with pale yellow spots or reticulations, occasionally plain or black, and a ventral surface white to yellowish. Tooth counts: 37–60/33–60. Vertebral counts: 86. Spiral valve counts: 12–14.

HABITAT AND RANGE: The Round Stingray is a benthic warm-temperate to tropical stingray usually found in nearshore waters less than 15 m deep, but reported down to a depth of at least 91 m. Round Stingrays prefer soft bottoms composed of mud or sand, often in areas where eelgrass, used for camouflage, is quite abundant. Water temperature plays an important role in the distribution of these stingrays as they prefer temperatures above a minimum of 50 degrees F. These rays are most abundant in the warm, shallow, coastal and bay waters of southern California from spring to fall. In winter they move into slightly deeper water where the temperature is more stable and less influenced by air temperature. Adults are more tolerant of temperature changes than juveniles.

This species is endemic to the eastern North Pacific from Humboldt Bay in northern California to Panama in Central America. It is most common off southern California and Baja, moving north of Point Conception only during periods of unusually warm coastal waters such as those coinciding with El Niño events.

NATURAL HISTORY: Viviparous, without a yolk sac placenta, with litters from one to six, with two or three being average. The litter size increases in larger females. Males and females both mature at an average disc width of 15 cm. Males may grow to a maximum disc width of 25 cm and females to a maximum disc width of 31 cm. The disc width at birth is 6 to 8 cm. These rays have evolved a combination of characteristics that significantly enhances their reproductive success. The ability of females to store sperm throughout the year provides the flexibility of two breeding seasons per year. Gestation is only three months long with most females giving birth every year. Maturity in both sexes is reached in about 2.6 years (31 months). Round Stingrays grow approximately 3 cm per year until they reach maturity, at which time their growth slows.

Round Stingrays show a strong segregation by age and sex. Adult females tend to live offshore in water deeper than 14 m, whereas adult males and juveniles tend to occupy a shallower habitat. Adult females move inshore to mate and give birth during spring and summer as the water temperature warms up. In southern California, mating occurs during spring, from April to June, with the young being born between June and October. Farther to the south, in the Gulf of California, mating occurs slightly earlier, from late winter to spring, with the pups being born three months later. A portion of the population mates and gives birth during winter, allowing for a second breeding season each year. After birth the pups remain in relatively shallow water (less than 2 m deep) until the onset of maturation. Adult females meanwhile return to deeper water after pupping and remain there until the following breeding season.

The feeding habits of Round Stingrays change as they grow in size. Juveniles from birth to about 14 cm in disc width consume mainly polychaete worms and small benthic crabs. As they reach adolescence and begin to mature, their diet shifts more to bivalve molluscs, crabs, and, to a lesser extent, polychaete worms. Adults feed prominently on bivalve molluscs and polychaete worms, with crabs making up a smaller proportion of their overall diet. This change in diet and preferred prey species is an evolutionary adaptation that reduces competition for food between adults and juveniles. The level of foraging activity among Round Stingrays coincides with seasonal changes in water temperature. Their feeding rate increases as the water temperature rises in summer and fall and decreases as the water temperature falls in winter and spring. They are voracious predators, foraging during daylight hours using their mouth and disc to dig large pits to get at buried prey items. The pits left by these rays appear to be an important process in structuring the resident benthic community in many areas. Once the Round Stingray has finished feeding, small fishes opportunistically feed on prey items in these pits that they ordinarily could not obtain on their own.

Round Stingrays are preyed upon by large sharks.

HUMAN INTERACTIONS: Round Stingrays are taken as a by-catch in commercial and recreational fisheries but are of no commercial value in California waters. To the south in Mexican waters they are considered more of a nuisance, as large groups of these rays often become entangled in shrimp nets.

Every year hundreds of beachgoers are accidently stung by these rays along the beaches of southern California. The most notorious of these beaches is an area at the northern end of Seal Beach that has been dubbed "Ray Bay." At this beach at least 474 people were treated for wounds from these stingrays between April and November 1952, and more than 500 bathers were stung during a ten-week period in 1962. During the 1990s on average 226 bathers were stung per year by these rays at Seal Beach, with the numbers higher in El Niño years. Stings are inflicted when bathers accidently step on this shallow water species. Although not fatal, these wounds can be quite painful. In recent years scientists working with local authorities have been capturing, removing the spines from, and tagging these rays in an attempt to reduce the number of bathers stung every year. Removal of the spine does not injure the ray as the spine will grow back by the following year.

NOMENCLATURE: *Urobatis halleri* (Cooper, 1863). The generic name comes from the Greek *uro,* meaning tail, and the Latin *batis,* meaning ray. The species is named after the young son of Major G. O. Haller, U.S. Army, who was stung on the foot while wading along the muddy shores of San Diego Bay. The common name refers to its nearly round disc.

The Round Stingray was originally placed in the genus *Urolophus*, but was later moved by Garman (1913) into his newly erected genus *Urobatis*, which included all eastern Pacific and western North Atlantic Round Stingrays. The genus *Urolophus* is restricted to those stingrays found in the western Pacific. Both genera are frequently seen in the literature for eastern Pacific forms, but current research suggests that *Urobatis* is the correct genus.

REFERENCES: Babel (1967); Nordell (1994); Valadez-Gonzalez et al. (2001); Van Blaricom (1982).

Whiptail Stingrays (Dasyatidae)

The whiptail rays are the largest stingray family, comprising six genera with between 66 and 77 species worldwide. One genera, with two species, is found in California waters, and a third species, the Longtail Stingray (*Dasyatis longus*), known from central Baja, would not be unexpected during extremely warm-water years. The whiptail rays, with one exception, are mostly benthic tropical and warm-temperate water rays. Most whiptail rays are found on the continental shelf to depths of at least 100 m, with some species residing in estuaries of reduced salinity and even in fresh water. A few species found in west Africa are actually restricted to a freshwater habitat, and one species leads an exclusively pelagic lifestyle. Their diet consists of benthic invertebrates and small bony fishes.

1a Snout pointed; brown to grayish or blackish above, white below. . . . . Diamond Stingray (*Dasyatis dipterura*)
1b Snout rounded; uniformly dark purple above and below . . . . Pelagic Stingray (*Dasyatis violacea*)

DIAMOND STINGRAY

Dasyatis dipterura

DESCRIPTION: The Diamond Stingray has a diamond-shaped disc that is slightly wider than it is long; a pointed snout; small molarlike teeth; a long, slender stinging spine; and a long whiplike tail less than 1.5 times the disc length. The dorsal surface is brownish to gray or black and reddish near the disc edges; the ventral surface is white. Vertebral counts: 105–124. Tooth counts: 21–37/23–44. Spiral valve count: 22–24.

HABITAT AND RANGE: The Diamond Ray is a warm-temperate to tropical stingray usually found on muddy or sandy bottoms near rocky reefs and kelp forests, down to a depth of at least 17 m. During the warm summer months Diamond Stingrays tend to congregate at depths of 2 to 7 m, but in fall and winter they move into deeper water usually between 12 and 17 m. Also during winter, for reasons that are unclear, they tend to concentrate around kelp beds more than on flat sandy bottoms.

Diamond Stingrays are found in the eastern Pacific ranging from southern California, where they are rare, to northern Chile and the Galapagos Islands. These stingrays are very common along the Pacific coast of Baja and in the Gulf of California. These rays are found in California waters more frequently and in greater abundance during periods coinciding with exceptionally warm water masses resulting from El Niño events. Reports of this species from British Columbian waters are unconfirmed. This species or a very similar looking one is found in Hawaiian waters, but its taxonomic status is uncertain.

NATURAL HISTORY: Viviparous, without a yolk sac placenta, with a litter size of one to four. Males mature at 50 to 65 cm with a maximum disc width of 86 cm. At maturity the disc width of females is about 65 cm, growing to a maximum of at least 88 cm and possibly up to 120 cm. The gestation period is about two to three months with the disc width at birth between 19 and 23 cm. Males and females tend to segregate by sex and age. Several bays along the Pacific coast of Baja are important nursery areas for Diamond Stingrays. Birth occurs in late summer, from August to September. Water temperature appears to play a role in the timing of birth. During El Niño years pupping may occur earlier in summer.

Diamond Stingrays feed mainly on benthic crustaceans, including crabs, shrimps, and small fishes. Diamond Stingrays, similar to other species of whiptail rays, forage by gliding just off the bottom searching for prey items. Once a prey item has been located the ray quickly settles on it; with a rapid, continuous up and down motion, using its pectoral fins for leverage, it then creates a vacuum to extract the prey item from its burrow. Once extracted the prey is seized in the ray's jaw and quickly devoured. These stingrays are most active at night as they forage for food, either singly or, more commonly, in groups of a few to hundreds.

HUMAN INTERACTIONS: Diamond Stingrays are too rare in California waters to be of any commercial importance. In Mexican waters this ray is abundant and of considerable commercial importance, comprising over 10 percent of the elasmobranchs caught seasonally. It is taken throughout the year with most of the effort coming during spring and summer when birthing occurs. It is also commonly taken by commercial fishers in Central and South American waters. Its wings are sold fresh or filleted and salted.

The long stinging spine of this ray makes it potentially dangerous to humans, and at least one death has been attributed to it. These rays usually flee quite rapidly when approached by divers.

NOMENCLATURE: *Dasyatis dipterura* (Jordan and Gilbert, 1880). The generic name is derived from the Greek *dasys,* meaning rough, and *batis,* meaning skate. The generic name is an abbreviation of its proper spelling, *Dasybatis.* The species name comes from the Latin *di,* meaning two, *ptero,* meaning wing, and *ura,* meaning tail. The common name is in reference to its diamond-shaped disc. It was once referred to as the Rat-Tailed Stingray.

The names *D. brevis* and *D. dipterura* are both frequently cited as the Diamond Stingray's scientific name. This confusion stems from its having been described by different authors in the same year. Jordan and Gilbert's description of *Dasybatis dipterura* was published in May 1880 based on a specimen from San Diego Bay, whereas Garman's original description of *D. brevis,* published in October of that year, was based on specimens collected both from San Diego, California, and Payta, Peru. Garman (1913) later synonymized these two species, assigning seniority to *D. brevis.* Because Jordan and Gilbert's *D. dipterura* was published first, it takes precedence over Garman's *D. brevis.*

REFERENCES: Mathews and Gonzalez (1975).

PELAGIC STINGRAY   *Dasyatis violacea*

DESCRIPTION: The Pelagic Stingray is characterized by a broad, wedge-shaped disc, width greater than length; a broadly rounded anterior margin and snout; teeth with a single pointed cusp, and sexually dimorphic, with the teeth of the male being longer and sharper; an extremely long stinging spine; and a whiplike tail longer than its disc length. It is a uniform dark purple above and below, although slightly lighter to lead gray below. Some literature accounts describe a thick black mucus covering the body. However, this mucus is exuded as a result of stress due to handling when captured. Tooth counts: 25–34/25–31. Vertebral counts: 94–99. Spiral valve counts: 18–22.

HABITAT AND RANGE: Among stingrays, Pelagic Stingrays are unique, preferring an oceanic environment to the seafloor bottom. They are found from the upper surface to a depth of at least 238 m over very deep water in tropical and warm-temperate regions. Pelagic Stingrays migrate north or south to higher latitudes following the movement of warm water masses but retreat as the water cools. In addition to migrating north or south they exhibit an inshore movement pattern, often moving over the edge of outer continental shelves seasonally as the water warms. These rays prefer water temperatures above 66 degrees F. Pelagic Stingrays have been recorded year-round in California waters.

Pelagic Stingrays range from British Columbia to Baja California and southward to central Chile and Easter Island, although they are uncommon north of Monterey Bay. There appear to be at least two discrete populations in the eastern North Pacific: one migrates from eastern Pacific equatorial waters to off the California coast, and a second central Pacific, population migrates northward occasionally winding up in Japanese and British Columbian waters. Although not reported from California waters until 1959, Pelagic Stingrays appear to be fairly commonplace. Elsewhere they are circumglobal in tropical and warm-temperate seas.

NATURAL HISTORY: Viviparous, without a yolk sac placenta, with litters of 4 to 13. Males mature at a disc width of 35 to 41 cm and grow to at least 59 cm disc width, although males held in public aquariums have grown to a maximum of 68 cm disc width. Disc width for females at maturity is 40 to 50 cm, with a maximum width of 80 cm and possibly up to 96 cm for individual specimens held in captivity. Disc width at birth is between 15 and 25 cm. Females are able to retain sperm for at least one year or more, thus allowing for some flexibility in their reproductive cycle if conditions are less than optimal. It is uncertain when mating takes place, although the gestation period for females, once fertilized, lasts about two to three months. In the eastern Pacific, births occur in winter off the coast of Central America in an area that also serves as a nursery ground.

Males mature at about two years and live at least five to six years, upto seven years in captivity. Females mature at three years and live at least seven to eight years, and up to nine years in captivity. Growth rates for captive males average 12.0 cm per year and for females 19.5 cm per year. These growth rates fluctuate seasonally, usually coinciding with feeding rate, with periods of rapid growth followed by a slower period.

Pelagic Stingrays feed on a variety of pelagic prey items including jellyfishes, crustaceans, squids, and small bony fishes. Captive specimens in public aquariums tend to be highly aggressive toward Giant Ocean Sunfishes, repeatedly biting and harassing them, especially if they are not fed. It is unknown whether this same behavior takes place in nature. Juveniles in captivity are voracious feeders consuming six to seven percent of their body weight per day, with the rate decreasing to a little more than one percent per day as they mature. Although active predators, their body coloration probably plays an important role in allowing them to approach equally active-swimming prey species. It also may camouflage them from potential predators such as sharks or toothed whales. Great White Sharks are known predators on Pelagic Stingrays.

HUMAN INTERACTIONS: The Pelagic Stingray is of no economic value but is frequently taken as by-catch in commercial fisheries.

The long, serrated stinging spine is potentially dangerous and may inflict serious wounds. At least two fatalities have been attributed to this species. A crewman aboard a long-line tuna boat was impaled by the spine when a ray was taken aboard and died within minutes. In another instance a crewman died from tetanus several days after being stung by one of these rays.

NOMENCLATURE: *Dasyatis violacea* (Bonaparte, 1832). The specific name comes from the Latin, *viola,* meaning purple, in reference to its body color. Its common name refers to its pelagic habitat.

The Pelagic Stingray is sometimes placed in the genus *Pteroplatytrygon* because of its unique pelagic habitat. However, current taxonomic studies reveal no significant morphological differences warranting placement of this ray in its own genus. Although Pelagic Stingrays have been described from other areas under various names, they all appear to be referable to a single wide-ranging species.

REFERENCES: Radovich (1961); Mollet (2002); Mollet et al. (2002).

Butterfly Rays (Gymnuridae)

The butterfly rays are a small group consisting of two genera and 12 species. A single species is known from California waters. These unmistakable rays have a broad, extremely flattened disc that is nearly twice as wide as it is long, a short tail, and a small stinging spine near the tail base. Maximum disc width ranges from 0.5 to at least 2 m. Butterfly rays are typically found in warm-temperate and tropical seas. All are nearshore bottom dwellers usually found in water less than 100 m deep. Development is viviparous, without a yolk sac placenta, and with litters of one to 14. Butterfly rays feed mainly on small bony fishes, and to a lesser degree on crustaceans and molluscs. They are of considerable commercial importance in Mexican waters and other regions.

CALIFORNIA BUTTERFLY RAY   *Gymnura marmorata*

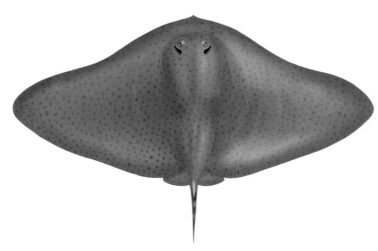

DESCRIPTION: These are the only rays in California waters with a broad disc nearly twice as wide as long; small pointed teeth; a small slender stinging spine; and a very short tail. They are brown to olive brown or grayish above, with a variable pattern of darker and lighter spots, and white below. Vertebral counts: 129. Tooth counts: 40–50/40–50. Spiral valve counts: 10–12.

HABITAT AND RANGE: The California Butterfly Ray is a common, nearshore ray usually found along sandy beaches and in shallow lagoons in warm-temperate and tropical waters.

Endemic to the eastern Pacific, these rays range from southern California to the Gulf of California, and southward to Peru. They are rare in California waters, except for periods of exceptionally warm water, but are extremely common to the south in the warm waters off Baja.

NATURAL HISTORY: Viviparous, without a yolk sac placenta, with litters of four to 16. The number of embryos increases with the size of the female. Disc width at maturity for males is 33 to 45 cm and for females about 68 cm. The maximum size for this species is 92 cm disc width for males and 122 cm disc width for females. Disc width at birth is 14 to 26 cm. Birth occurs in May and June.

California Butterfly Rays feed mainly on bony fishes and on rare occasion on squid. These rays capture their prey by stunning it with their "wings" and then using them to manipulate the prey item into their mouths. Some of the fish species they consume are nearly as long as their disc length.

HUMAN INTERACTIONS: The California Butterfly Ray is occasionally caught by commercial and recreational fishermen but is of no fishery importance in California waters. However, in Mexican waters, particularly in the Gulf of California, this is one of the most important elasmobranch species taken commercially. Its white flesh is favored over the flesh of other rays in Mexican waters.

The California Butterfly Ray, unlike the Round Stingray, has a short stinging spine that poses little danger to bathers. These docile, harmless rays are occasionally seen swimming by divers. They are quite impressive rays to observe swimming, with a butterfly-like undulatory motion of their enormous pectoral fins.

NOMENCLATURE: *Gymnura marmorata* (Cooper, 1864). The generic name comes from the Greek *gymnos,* meaning bare. The species name comes from the Latin *marmoratus,* meaning marbled. The common name refers to the locality where it was first described and to its broad butterfly-like shape. They are also referred to as the Butterfly Stingray.

The California Butterfly Ray was originally described by Cooper (1864) as *Pteroplatea marmorata* from a specimen caught off San Diego. The genus *Pteroplatea* remained in common usage locally until Fowler (1941) synonymized it with the genus *Gymnura*. The Mazatlan Butterfly Ray (*G. crebripunctata*) was based on a description of an adult male California Butterfly Ray, thus relegating the former to junior synonym status.

REFERENCES: Villavicencio-Garayzar (1993b).

Eagle Rays (Myliobatidae)

The eagle rays are a moderate-sized family with four genera and 18 to 21 species. A single species occurs in California waters, although a second species, the Longnose Eagle Ray (*Myliobatis longirostris*), known from the Pacific coast of southern Baja, might be found during periods of exceptionally warm water. These are medium to large stingrays with a laterally expanded disc more than 1.5 times its length; a short, thick, bluntly rounded snout; a relatively small, transverse mouth with large, flattened platelike teeth for crushing; and a long whiplike tail. Some species may have a disc width of up to 2.4 m or more, but most are less than 2 m. Eagle rays are common inhabitants of temperate and tropical seas. They are usually found on continental shelves from shallow nearshore waters, including bays and sloughs, to a depth of over 100 m, often migrating over very deep water moving between the continental mainland and offshore islands. Eagles rays will often form large aggregations when foraging for food, mating, or making long-distance migrations. These rays do well in captivity and are often on display in public aquariums. In some regions they are among the most important elasmobranchs taken in commercial fisheries.

BAT RAY *Myliobatis californicus*

DESCRIPTION: The Bat Ray is a heavy-bodied eagle ray with a disc width more than 1.5 times its length; a massive head slightly raised above the disc; platelike teeth; a slender, elongated, finely serrated sting- ing spine; and a long, slender, whiplike tail. The dorsal surface is dark brown, olive, or black, and the ventral surface is white. Albinism has been reported in this species. Vertebral counts: 82. Spiral valve counts: 16–19.

HABITAT AND RANGE: Bat Rays are a cool to warm-temperate species commonly found in bays and sloughs along the mainland and around the Channel Islands. They are common on mud flats and sandy bottoms, as well as around rocky reefs and kelp forests, usually in water less than 50 m deep, although they will migrate over considerably deeper water moving between offshore islands and the mainland.

They are endemic to the eastern Pacific, ranging from Yaquina Bay in northern Oregon to the Gulf of California.

NATURAL HISTORY: Viviparous, without a yolk sac placenta, with litters of two to 12. Litter size increases from two to four in smaller females up to 10 to 12 in larger individuals. The pups are born during late spring and summer after a nine- to 12-month gestation period. Males mature at a disc width of 62 to 66 cm and grow to a maximum disc width of 101 cm. Disc width for females at maturity is 88 to 100 cm with a maximum of 180 cm, although specimens over 150 cm are uncommon. The size at maturity varies slightly between populations located in northern California and those from off Baja suggesting that discrete subpopulations of these rays are found along the coast. Disc width at birth is 20 to 31 cm.

Several bays and sloughs along the California and Pacific Coast of Baja are important nursery and feeding grounds for Bat Rays, which are seasonally abundant from spring through fall. Adult females are most abundant during spring and early summer when they give birth and outnumber adult males during this time by ratios of up to 6:1. This imbalance in the sex ratio of adults evens out, however, during summer when large mating aggregations of these rays gather within the confines of these same coastal bays and sloughs. After the summer mating season, adult Bat Rays remain within these bays until early fall, at which time they move out into open coastal waters. Juveniles reside within these food-enriched bays most of the year, leaving only in late fall and winter when the water temperature drops below 50 degrees F but returning as it warms in spring.

Individual Bat Rays seasonally return to the same bays and sloughs to give birth and mate, providing further evidence that each of these areas represents a discrete subpopulation along the coast. Very little is known about their movements outside of bays. Large schools of similar-sized Bat Rays often numbering into the thousands have been seen around the Channel Islands, along the mainland coast, and in bays, but little is known as to why these aggregations form. Bat Rays have been observed to school with Brown Smoothhound and Leopard Sharks in northern California and with the Spotted Eagle Ray off Baja.

Males mature at about two to three years and live to a maximum age of at least six years. Females mature at about five years and live to at least 24 years. Newborns appear to have a fairly rapid growth rate during the first one to two years of life, growing from 10 to 20 cm annually, but somewhat more slowly after maturity has been attained.

Bat Rays feed mainly on benthic invertebrates including abalone, clams, snails, crabs, shrimps, echiuroid worms, sea cucumbers, brittle stars, polychaete worms, and, on rare occasion, small bony fishes. Interestingly, the prey preference changes as these rays grow in size and reach maturity. Juveniles feed mainly on small clams, and, to a lesser extent, on polychaete worms and shrimps. Larger adolescent Bat Rays show a preference for crabs as well as clams. Adults feed on large clams and echiuroid worms that younger rays are unable to dig up. The adults obtain these deeper burrowing prey items by digging pits, sometime measuring up to 20 cm deep and 400 cm wide. However, females, which generally grow larger, can dig deeper pits to get at prey species that smaller adult males cannot reach. This change in diet from juvenile to adult is an important evolutionary strategy that ensures a plentiful food supply for younger rays, which do not have to compete with larger, more active adults for the same food resource.

Adult Bat Rays are quite powerful, and their digging action can modify the bottom. The pits left by these rays provide an important food resource to other species, particularly flatfishes, such as sand dabs, and horn sharks, which will feed in these pits after the Bat Ray has finished. The exposed prey items would ordinarily be unobtainable by these fishes, which lack the Bat Ray's powerful digging ability. Juvenile Horn Sharks also use these pits as camouflage in which to hide from potential predators.

Predators on Bat Rays include Great White Sharks and Sevengill Sharks. Sevengills are especially voracious predators on Bat Rays in Humboldt and San Francisco Bays. Juvenile Sevengills will hunt relatively large Bat Rays in packs, with five or six Sevengills cooperatively attacking, killing, and consuming the ray. California Sea Lions will occasionally feed on juvenile Bat Rays.

HUMAN INTERACTIONS: Bat Rays are taken in considerable numbers by recreational anglers, particularly in Elkhorn Slough and San Francisco Bay, who often target them because of their active fighting ability when hooked. The number of Bat Rays caught and the potential impact on their population are largely unknown as catch data for recreational fisheries are lacking for most elasmobranch species in California waters. There has never been a directed commercial fishery in California, although they are taken commercially in Mexican waters as food for human consumption. The flesh of their pectoral fins is quite tasty.

For many years Bat Rays were believed to be destructive to oyster beds and were subject to predator control measures to reduce their population. Detailed studies on their feeding habits, however, revealed that although they are major predators on clams and crabs, oysters are rarely eaten, and then only by larger females. An ironic twist is that crabs, which are heavily preyed upon by Bat Rays, are in fact major predators on oysters, and thus the reduction of these rays by oyster growers has inadvertently affected their crop by allowing oyster-eating crabs to proliferate.

Bat Rays are fairly docile and will quickly take flight if disturbed or approached by divers. Anglers catching these rays should be wary of their stinging spine at the base of their tail and their powerful crushing jaws.

Bat Rays are maintained in display exhibits at many public aquariums. They are fairly intelligent rays and can be hand-fed by aquarium visitors.

NOMENCLATURE: *Myliobatis californicus* (Gill, 1865). The generic name comes from the Greek *mylios*, meaning grinder, and the Latin *batis*, meaning ray. The species name refers to the region in which the original specimen was described. The common name is in reference to its batlike shape. Other local names include Bat Fish, Bat Stingray, California Stingray, and Eagle Ray.

The Bat Ray has a somewhat confusing taxonomic history. It was originally described as *Rhinoptera vespertilio* by Girard (1856) based on a specimen from Tomales Bay. However, the species name *vespertilio* was invalidated as it was already occupied by the Ornate Eagle Ray, *Myliobatis vespertilio* (= *Aetomylaeus vespertilio*), a western Indo-Pacific ray that had previously been described by Bleeker (1852). Gill (1862) considered the Bat Ray distinct from the genus *Myliobatis* and erected the genus *Holorhinus* placing Girard's *vespertilio* in it. Subsequent taxonomic research revealed that the proper genus was indeed *Myliobatis* and thus Gill (1865) redescribed it as *M. californicus*. The species name *Holorhinus californicus*, however, remained in common usage for many years and is still occasionally seen in the literature.

REFERENCES: Gray et al. (1997); Martin and Cailliet (1988a,b); Matern et al. (2000); Van Blaricom (1982); Villavicencio-Garayzar (1995b).

Devil Rays (Mobulidae)

The devil rays consist of two genera and 10 species, with both genera and two species being represented in California waters. The devil rays are easily distinguished from all other rays by their broad winglike body. These are among some of the largest known rays, with some species exceeding a disc width of 6 m. Devil rays are majestic swimmers that may be solitary or may be seen in small aggregations of 30 to 50 individuals or more. These are warm-temperate to tropical oceanic rays found from close inshore to far out at sea. They are occasionally seen in California waters, particularly during extremely warm-water years.

1a   Mouth on front of head; teeth on lower jaw only . . . .
. . . . Manta Ray (*Manta birostris*)

1b   Mouth on underside of head; teeth on both jaws . . . .
. . . . Spinetail Mobula (*Mobula japanica*)

MANTA FOLLOWS ➤

MANTA

Manta birostris

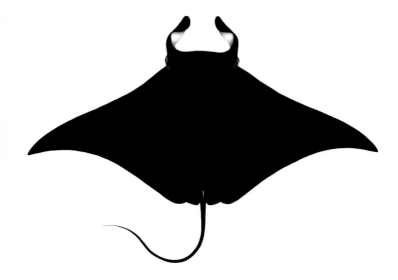

DESCRIPTION: The Manta is an enormous devil ray with a large, broad head; prominent hornlike flaps; a terminal mouth; teeth rounded to square, small, and present only on the lower jaw; a very small sting- ing spine sometimes appearing as little more than a knob behind the dorsal fin; and a whiplike tail shorter than the disc length. It is black to dark brown above, often with white shoulder patches, white below, but becoming grayish along the edges of the fin. A unique color morph known as a Black Manta is completely dark below. Albinism has been reported in this species. Individual Mantas can be recognized by their color patterns, especially on their abdomen. They are able to vary the color pattern on their pectoral fins with pale patches appearing at night, but becoming uniformly darker during the day. The significance of this daily color change is unknown. Tooth counts: —/96–360. Spiral valve counts: 40–45.

HABITAT AND RANGE: Mantas are found in tropical to warm-temperate areas in which the water temperature is usually above 68 degrees F. They follow the movement of warm water masses into California water but retreat as the water cools. A coastal pelagic to oceanic species, they are usually seen at the surface over continental shelf waters and around offshore islands. Mantas prefer areas of rich biological activity such as upwelling zones. They may be seen swimming singly or in groups of 30 to 50, sometimes in association with other sharks and rays, large bony fishes, marine mammals, and birds. Remoras are commonly seen clinging to Mantas, especially on the anterior parts of their pectoral fins and on their ventral surface. The swimming behavior of Mantas changes in relation to their habitat. Those traveling off the continental shelf over deep water (greater than 200 m) are usually swimming at a sustained rate with a clear intended direction, whereas those close inshore (less than 200 m) are usually either basking or slowly swimming without a clear direction or purpose. Mantas may be seen year-round in coastal regions but are most abundant from spring to fall. In winter they seem to disperse either to deeper water, further offshore, or to other locations.

In California waters Mantas are found as far north as Santa Barbara and around the Channel Islands in southern California. They are very common off Baja and throughout the tropical eastern Pacific to Peru and the Galapagos Islands. Elsewhere, Mantas are found worldwide in most warm-temperate and tropical seas.

NATURAL HISTORY: Viviparous, without a yolk sac placenta, with litters of one or two, although most have only a single pup per litter. Mating occurs in late spring and early summer with birth taking place one to two years later. It is unclear whether these rays have an annual or biennial reproductive cycle. Birth takes place during summer in the eastern Pacific, with the disc width at birth between 1.1 and 1.3 m. Mantas exhibit an unusual birthing behavior whereby the female breaches the water's surface and ejects the young into the air. Males mature at a disc width of about 4 m and females at about 4.3 m. Mantas are the largest living rays and one of the largest living fishes, with a maximum wing span of at least 6.7 m, with anecdotal reports of some being over 9 m.

Anecdotal evidence indicates that Mantas mature in about six years and live over 15 to 20 years, but this is unverified.

Mantas are filter feeders that consume large quantities of planktonic crabs, shrimps, and euphausiids, as well as small fishes such as mullet. A single Manta will consume about 13 percent of its body weight per week of plankton. This roughly translates into about 200 kg of plankton per week for a Manta with a 5 m disc width. Mantas feed by swimming in a slow circle to concentrate the plankton into a tight "bait-ball"; they then swim through it, directing the minute prey into their mouths using their hornlike flaps. The plankton is strained out by specialized gill filaments that retain the minute organisms. Mantas will migrate from one plankton-rich area to another, but appear to remain in the same localized area for most of the year or at least while their preferred prey is available. Mantas usually feed at or near the surface, but once they have finished feeding they will move near the bottom to "cleaning stations" so that small wrasses and cleaner shrimp can tend to them by removing small parasites from the Mantas.

During the day Mantas swim from near the surface to a depth of about 50 m but usually stay off the bottom. However, at night they swim close to the bottom at depths of 50 to 200 m with occasional trips to the surface. They tend to move closer to land during the day, coming into very shallow water, but at night will move farther offshore to depths of 100 to 200 m. The reasons for these daily movement patterns are unclear but may be related to feeding. Support for this observation comes from the fact that Mantas can be attracted inshore at night if building lights on shore are left on. The lights attract numerous planktonic species, which eventually bring in the Mantas.

Sharks, and possibly Killer Whales, are the most common predator on Mantas. A single Bull Shark was once observed to consume a Manta with a 2.5 m disc width. Bite marks typical of those made by the Cookiecutter Shark have also been observed. Very large Mantas probably have very few predators.

HUMAN INTERACTIONS: Mantas are occasionally taken as a bycatch in the drift gill net fishery, but they are too rare in California waters to be of any commercial value. In Baja waters they are caught seasonally in small numbers, and their wings are utilized as food for human consumption.

Old tales of Mantas devouring divers, either by clasping them with their hornlike flaps or enveloping them with their pectoral fins, are without foundation. In fact, in recent years a cottage ecotourist industry has developed in Baja Sur, and elsewhere, in which charter boats take divers out to dive with these gigantic, majestic rays. This is a dramatic change from a half century ago when boats would take people out to harpoon these rays for sport. Mantas are fairly docile rays and are easily approached; however, their size and power should not be underestimated as they have been known to capsize small boats with a single blow from their massive wings if they become entangled in anchor lines or are harpooned.

NOMENCLATURE: *Manta birostris* (Donndorff, 1798). The common and generic scientific name, Manta, is Spanish for blanket and is in reference to their enormous size. The species name comes from the Latin *bi,* meaning two, and *rostris* (= *rostrum*), meaning snout in reference to its two hornlike flaps. Other names include Blanket Fish, Devil Fish, Devil Ray, Manta Ray, and Sea Devil.

Atlantic and Pacific forms were at one time considered separate species, although most researchers now consider them to be a single wide-ranging species. The eastern Pacific form was once describe as *Manta hamiltoni,* and this name is commonly seen in earlier literature. Authorship for the species name is sometimes incorrectly ascribed to Walbaum (1792).

REFERENCES: Barton (1948); Homma et al. (1999); Walford (1931); Yano et al. (1999).

SPINETAIL MOBULA FOLLOWS ➤

SPINETAIL MOBULA

Mobula japanica

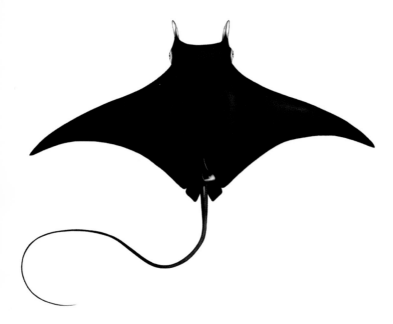

DESCRIPTION: The Spinetail Mobula is a medium-sized devil ray with sharply pointed pectoral fin apices; a subterminal mouth; small similar-looking teeth on the upper and lower jaws; a short, serrated, sharp stinging spine behind the dorsal fin;

and a very long wirelike tail equal to or slightly longer than its disc width. It is bright blue to black above; it is white below in juveniles, but becomes dotted with dark blotches in adults. Juveniles have white crescent-shaped shoulder bars, which usually fade in adults. Tooth counts: 95–295/112–304. Spiral valve counts: 51–57.

HABITAT AND RANGE: The Spinetail is a wide-ranging mobula found in warm-temperate and tropical waters. It is a regular visitor to California coastal waters during warm summer months and may be resident year-round when water temperatures are above normal. They are usually seen at the surface, along coastal waters and around offshore islands, swimming either singularly or in small groups.

In the eastern Pacific, Spinetail Mobulas range from central California, including the Channel Islands, to Peru. This species is most likely circumglobal in its distribution, although records from the Atlantic and Indian Oceans are sketchy.

NATURAL HISTORY: Viviparous, without a yolk sac placenta, with possibly only one pup being born at a time. Disc width at maturity for males and females is about 210 cm, with a maximum of 310 cm. Disc width at birth is between 70 and 85 cm. The gestation period may be one to two years or more. Mating takes place during June and July in the lower Gulf of California, but this area does not appear to serve as a nursery. In the eastern Pacific, females with developing embryos are found during spring with term embryos seen in late summer. Birthing appears to take place away from the mainland, possibly around offshore islands or seamounts.

The Spinetail Mobula is primarily a filter feeder, feeding on planktonic species such as euphausiid shrimps and copepods. These are highly active, migratory rays that will follow the daily and seasonal movement patterns of their primary prey species. The lower Gulf of California appears to serve as an important feeding area during spring and summer when euphausiid shrimp are abundant. As the abundance of their preferred prey declines in fall, so does the abundance of the Spinetail Mobula.

HUMAN INTERACTIONS: The Spinetail Mobula is not fished in California waters but is taken occasionally as a by-catch in other fisheries. In the lower Gulf of California they are seasonally taken in small numbers by commercial fisheries and the meat is used as food for human consumption. The Pygmy Mobula (*M. munkiana*) is a more important component of the elasmobranch fisheries in the lower Gulf of California.

NOMENCLATURE: *Mobula japanica* (Muller and Henle, 1841). The generic name *Mobula* is of uncertain origin, but may derive from the Latin *mobilis,* meaning movable, in reference to its migratory habits. The name could also possibly allude to furniture, which apparently in Romance languages is referred to as "movables" or nonfixed furnishings. Rafinesque (1810) mentions the Italian vernacular *Tavila cornuta* or "horned table" in his description of *Mobula mobular,* which suggests furniture. So a *Mobula* may be a "movable" because it looks like a piece of furniture such as a table, and possibly because it is mobile or highly active. If the generic name *Manta* can refer to a blanket, *Mobula* may refer to a flying table! The species name *japanica* is for the locality, Japan, in which it was first described. The common name refers to the small tail spine of this species. It has also been referred to as the Devil Ray or Mobula in California waters.

The species was originally described as *M. japanica*, which is its proper spelling. However, subsequent authors corrected this, opting to use the proper Latin spelling *japonica*. This latter species name is often used incorrectly in the literature. The first record of this species in California waters came from a specimen caught off Laguna Beach. It was initially misidentified as the Smooth-Tail Mobula (*M. lucasana*), a name that appears in some literature accounts, but a species that to date has not been found in California waters.

REFERENCES: MacGinitie (1947); Notarbartolo-di-Sciara (1987, 1988).

CHIMAERAS (CHIMAERIFORMES)

The chimaeras, or ratfishes, are a small, primitive group of chondrichthyans, all of which occur within a single order composed of three families, six genera, and at least 50 known species. The number of species within this group is likely to increase as at least 15 species are currently undescribed, including two from California waters. Two families and at least four species are known from California waters. The chimaeras are readily identified by their elongated tapering body; filamentous, whiplike tail; smooth scaleless skin; large venomous dorsal fin spine; winglike pectoral fins; and open lateral line canals, which appear as grooves on the head and along the trunk. Adults range in size from 60 to just over 150 cm as measured from their snout tip to the posterior edge of the caudal fin. In some species the tail extends beyond the caudal fin as an elongated terminal whip that can reach extreme lengths (sometimes equaling the body length). Although most chimaeras do not grow to lengths much over 1 m, some of the deepwater species, particularly in the families Chimaeridae and Callorhynchidae, grow to massive sizes by increasing their body diameter and overall bulk. Like elasmobranchs, chimaeras have internal fertilization, and all the males are equipped with pelvic claspers used to transfer sperm to the female. However, male chimaeras have taken clasper morphology to an extreme as each pelvic clasper may be divided into two, and sometimes three, fleshy denticulate rods. These are all egg-laying species, but very little is known about their reproductive biology as most live in waters more than 500 m deep.

1a Snout short and blunt . . . . shortnose chimaeras (Chimaeridae)
1b Snout elongated and tapering . . . . longnose chimaeras (Rhinochimaeridae)

Shortnose Chimaeras (Chimaeridae)

The shortnose chimaeras are the largest of the three chimaera families, with two genera, and at least 19 described species, with perhaps 15 or more species awaiting formal description. One genus with three species, two of which are undescribed, occurs in California waters. The family is distinguished by a short, blunt, fleshy snout. Worldwide in their distribution, shortnose chimaeras as a group are generally found in waters between 60 and 1,000 m deep, with some species occurring in shallower waters, even intertidally, and others being found at a depth of 2,500 m. They are commonly taken as a by-catch.

1a Body surface with numerous, scattered white spots. Second dorsal fin with a deep notch, giving it the appearance of two fins . . . . White-spotted Ratfish (*Hydrolagus colliei*)

1b Body surface without spots or other markings. Second dorsal fin continuous, without a notch . . . . Black Chimaera (*Hydrolagus* spp.)

WHITE-SPOTTED RATFISH *Hydrolagus colliei*

DESCRIPTION: The White-spotted Ratfish is a ratfish with a large head and a tapering body; a short, blunt snout; large, luminescent green eyes; a prominent dorsal fin spine that is slightly higher than the dorsal fin; and a long, low second dorsal fin with an undulating profile giving it the appearance of two fins. It is silvery gray or brownish with varying hues of gold, green, and blue; the upper surface has numerous white spots. The caudal and dorsal fins have darker edges.

HABITAT AND RANGE: White-spotted Ratfishes prefer areas of mud or cobblestone, with some vertical relief from boulders or reefs to which they orient. They are found from the intertidal zone to a depth of 971 m. They are common in bays and sounds to the north, but are found in progressively deeper water in the southern portion of their range. Off British Columbia they are found close inshore, even intertidally, but in southern California they are rare in water less than 30 m deep. They are common on some southern California reef slopes, particularly those off Malibu and Redondo Beach, at depths below 30 m. In the Gulf of California they are rare in water less than 183 m deep. They prefer water temperatures of 45 to 48 degrees F.

White-spotted Ratfishes appear to have discrete populations along the coast as they are quite abundant in some areas and sparse in others. Very little is known about their movement patterns other than they seem to be seasonally abundant in some areas. In Puget Sound they move inshore at night but retreat into deeper water during the day.

They range from southeast Alaska to Cedros Island, Baja, and the northern Gulf of California, being most common between British Columbia and southern California.

NATURAL HISTORY: Oviparous, they have spearlike egg cases that require 18 to 30 hours to expel during parturition. After being extruded the egg cases may hang freely in the water for four to six days, attached to the female by a long, slender extension of the egg case called an elastic capsular filament. Egg cases may be deposited in the mud erect or on a gravel substrate. They are produced throughout the year, with the maximum spawning period occurring during spring and summer. The incubation period within the egg case is about 12 months, with the newborns emerging at about 14 cm in length. They grow to about 30 cm during their first year of life. Males mature at 35 to 44 cm and females at 41 to 47 cm. Their maximum length is about 1 m. Maturity is reached in about four years.

White-spotted Ratfishes feed on a wide variety of benthic invertebrates including polychaete worms, molluscs, crustaceans, isopods, and echinoderms, as well as small benthic fishes. They are known to be cannibalistic, feeding on both free-swimming individuals as well as their own egg cases. White-spotted Ratfishes are feeble swimmers and tend to select prey items having a limited escape response. These ratfishes seek out their prey primarily by smell and electroreception.

White-spotted Ratfishes are an important prey source for several diverse species including the Sixgill Shark, Sevengill Shark, Soupfin Shark, Spiny Dogfish, Giant Sea Bass, Lingcod, rockfish, Pacific Halibut, and marine mammals.

HUMAN INTERACTIONS: White-spotted Ratfishes are taken as a by-catch but are of no commercial value in California waters.

Great care should be taken when handling the White-spotted Ratfish as its spine is venomous and it can inflict a very painful injury if mishandled.

NOMENCLATURE: *Hydrolagus colliei* (Lay and Bennett, 1839). The generic name comes from the Greek *(h)ydro,* meaning water, and *lagos,* meaning hare, in reference to its rabbitlike head. The species name is after the naturalist Alexander Collie, who collected the type specimen in Monterey Bay during Captain Beechey's voyage of the *HMS Blossom*. The common name is in reference to the numerous scattered white spots covering its body. It is also referred to as the Ratfish or Chimaera.

The species was originally placed in the genus *Chimaera* but was subsequently moved when Gill (1862) erected the genus *Hydrolagus*.

REFERENCES: Johnson and Horton (1972); Kato (1992); Quinn et al. (1980).

BLACK CHIMAERA $\qquad$ *Hydrolagus* spp.

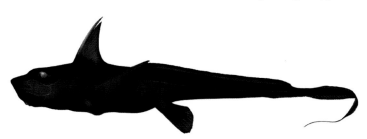

DESCRIPTION: The Black Chimaera has a relatively large head, with a short, blunt snout; large eyes; a dorsal fin spine that is slightly higher than the first dorsal fin; and a long, low, continuous second dorsal fin without a notch. It is uniformly bluish black to black, without any spots or other markings.

HABITAT AND RANGE: It is known from only a few specimens caught off southern California at a depth of 925

to 1,400 m. A few specimens were also taken in very deep water in the Gulf of California.

NATURAL HISTORY: Males reach a maximum size of 110 cm, but nothing else is known about its biology.

HUMAN INTERACTIONS: It is occasionally caught in Sablefish traps set in very deep water.

NOMENCLATURE: *Hydrolagus* spp. This species is currently being described. The common name is in reference to its overall dark body coloration. A second undescribed species has been video-taped on the Davidson Seamount off central California at a depth of over 2,000 m.

REFERENCES: Eschmeyer et al. (1983).

Longnose Chimaeras (Rhinochimaeridae)

The longnose chimaeras are composed of three genera and seven species. A single species occurs in California waters. These chimaeras all have an extremely elongated, fleshy snout. The function of the snout is uncertain, but the presence of numerous ampullary sensory organs suggests that it most likely aids in prey detection. Longnose chimaeras are found in all major oceans ranging at depths from 750 to 2,800 m. These are the least-studied group of chimaeras.

LONGNOSE CHIMAERA   *Harriotta raleighana*

DESCRIPTION: The Longnose is a chimaera with a very long, laterally expanded snout; a prominent first dorsal fin spine; and a long, low second dorsal fin. It is unlikely to be confused with any other chimaera currently known in California waters. It is uniformly chocolate brown with slightly darker fin edges; the pelvic fins are brownish black. Abrasions from capture in nets often leave pale areas on the skin.

HABITAT AND RANGE: A deepwater species, the Longnose Chimaera is usually found at a depth of 530 to 2,603 m.

The only eastern Pacific records are based on a few specimens taken between southern California and the Gulf of California. Elsewhere this species has a scattered circumglobal distribution.

NATURAL HISTORY: Oviparous, with distinctive egg cases that have broad lateral flanges with numerous transverse ridges. Males mature at about 73 cm. Female maturity is uncertain, but they do grow to at least 99 cm. The maximum size is 120 cm including its long filamentous tail. The size at birth is about 13 cm. Birth may occur during summer.

The Longnose Chimaera feeds mainly on polychaete worms, molluscs, and crustaceans.

HUMAN INTERACTIONS: The Longnose Chimaera has been taken incidentally in Sablefish traps from very deep water off southern California. Care should be exercised when handling this species as its spine can inflict a painful injury if it is mishandled.

NOMENCLATURE: *Harriotta raleighana* (Goode and Bean, 1895). The genus was named in honor of Thomas Harriott, a prominent English philosopher and naturalist and member of the Roanoke Colony, Virginia, in 1585, who wrote the first work in English on American natural history. The species is named in honor of Sir Walter Raleigh, who founded the Roanoke Colony in Virginia, and who is responsible for sending the first English scientific explorer to the New World. Its common name is in reference to its long, slender snout.

REFERENCES: Eschmeyer et al. (1983).

EGG CASES

Horn Shark
(*Heterodontus francisci*)

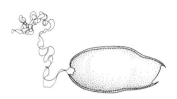

Brown Catshark
(*Apristurus brunneus*)

Swell Shark
(*Cephaloscyllium ventriosum*)

Longnose Catshark
(*Apristurus kampae*)

Filetail Catshark
(*Parmaturus xaniurus*)

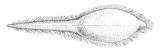

White-spotted Chimaera
(*Hydrolagus colliei*)

Longnose Chimaera
(*Harriotta raleighana*)

Aleutian Skate
(*Bathyraja aleutica*)

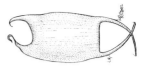

Sandpaper Skate
(*Bathyraja kincaidii*)

Pacific White Skate
(*Bathyraja spinosissima*)

Roughtail Skate
(*Bathyraja trachura*)

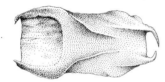

Big Skate
(*Raja binoculata*)

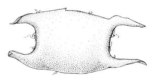

California Skate
(*Raja inornata*)

Longnose Skate
(*Raja rhina*)

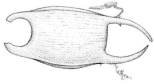

Pacific Starry Skate
(*Raja stellulata*)

CHECKLIST OF CALIFORNIA CHONDRICHTHYANS

Cow and Frilled Sharks (Hexanchiformes)

FRILLED SHARKS (CHLAMYDOSELACHIDAE)
☐ Frilled Shark *(Chlamydoselachus anguineus)*

SIXGILL AND SEVENGILL SHARKS (HEXANCHIDAE)
☐ Sixgill Shark *(Hexanchus griseus)*
☐ Sevengill Shark *(Notorynchus cepedianus)*

Dogfish Sharks (Squaliformes)

BRAMBLE SHARKS (ECHINORHINIDAE)
☐ Prickly Shark *(Echinorhinus cookei)*

DOGFISH SHARKS (SQUALIDAE)
☐ Spiny Dogfish *(Squalus acanthias)*

LANTERNSHARKS (ETMOPTERIDAE)
☐ Pacific Black Dogfish *(Centroscyllium nigrum)*

SLEEPER SHARKS (SOMNIOSIDAE)
☐ Pacific Sleeper Shark *(Somniosus pacificus)*

KITEFIN SHARKS (DALATIIDAE)
☐ Pygmy Shark *(Euprotomicrus bispinatus)*
☐ Cookiecutter Shark *(Isistius brasiliensis)*

Angel Sharks (Squatiniformes)

ANGEL SHARKS (SQUATINIDAE)
☐ Pacific Angel Shark *(Squatina californica)*

Horn Sharks (Heterodontiformes)

HORN SHARKS (HETERODONTIDAE)
- [ ] Horn Shark *(Heterodontus francisci)*

Carpet Sharks (Orectolobiformes)

WHALE SHARKS (RHINCODONTIDAE)
- [ ] Whale Shark *(Rhincodon typus)*

Mackerel Sharks (Lamniformes)

SAND TIGER SHARKS (ODONTASPIDIDAE)
- [ ] Sand Tiger Shark *(Odontaspis ferox)*

GOBLIN SHARKS (MITSUKURINIDAE)
- [ ] Goblin Shark *(Mitsukurina owstoni)*

MEGAMOUTH SHARKS (MEGACHASMIDAE)
- [ ] Megamouth Shark *(Megachasma pelagios)*

THRESHER SHARKS (ALOPIIDAE)
- [ ] Pelagic Thresher Shark *(Alopias pelagicus)*
- [ ] Bigeye Thresher Shark *(Alopias superciliosus)*
- [ ] Common Thresher Shark *(Alopias vulpinus)*

BASKING SHARKS (CETORHINIDAE)
- [ ] Basking Shark *(Cetorhinus maximus)*

MACKEREL SHARKS (LAMNIDAE)
- [ ] Great White Shark *(Carcharodon carcharias)*
- [ ] Shortfin Mako Shark *(Isurus oxyrinchus)*
- [ ] Longfin Mako Shark *(Isurus paucus)*
- [ ] Salmon Shark *(Lamna ditropis)*

Ground Sharks (Carcharhiniformes)

CATSHARKS (SCYLIORHINIDAE)
- [ ] Brown Catshark *(Apristurus brunneus)*
- [ ] Longnose Catshark *(Apristurus kampae)*
- [ ] Swell Shark *(Cephaloscyllium ventriosum)*
- [ ] Filetail Catshark *(Parmaturus xaniurus)*

HOUNDSHARKS (TRIAKIDAE)
- [ ] Soupfin Shark *(Galeorhinus galeus)*
- [ ] Gray Smoothhound Shark *(Mustelus californica)*

☐ Brown Smoothhound Shark *(Mustelus henlei)*
☐ Sicklefin Smoothhound Shark *(Mustelus lunulatus*
☐ Leopard Shark *(Triakis semifasciata)*

REQUIEM SHARKS (CARCHARHINIDAE)
☐ Copper Shark *(Carcharhinus brachyurus)*
☐ Bull Shark *(Carcharhinus leucas)*
☐ Blacktip Shark *(Carcharhinus limbatus)*
☐ Oceanic Whitetip Shark *(Carcharhinus longimanus)*
☐ Dusky Shark *(Carcharhinus obscurus)*
☐ Tiger Shark *(Galeocerdo cuvier)*
☐ Blue Shark *(Prionace glauca)*
☐ Pacific Sharpnose Shark *(Rhizoprionodon longurio)*

HAMMERHEAD SHARKS (SPHYRNIDAE)
☐ Scalloped Hammerhead Shark *(Sphyrna lewini)*
☐ Bonnethead Shark *(Sphyrna tiburo)*
☐ Smooth Hammerhead Shark *(Sphyrna zygaena)*

Rays (Rajiformes)

GUITARFISHES (RHINOBATIDAE)
☐ Shovelnose Guitarfish *(Rhinobatos productus)*
☐ Banded Guitarfish *(Zapteryx exasperata)*

THORNBACK RAYS (PLATYRHINIDAE)
☐ Thornback Ray *(Platyrhinoidis triseriata)*

TORPEDO RAYS (TORPEDINIDAE)
☐ Pacific Torpedo Ray *(Torpedo californica)*

SOFTNOSE SKATES (ARHYNCHOBATIDAE)
☐ Deepsea Skate *(Bathyraja abyssicola)*
☐ Aleutian Skate *(Bathyraja aleutica)*
☐ Sandpaper Skate *(Bathyraja kincaidii)*
☐ Fine-spined Skate *(Bathyraja microtrachys)*
☐ Pacific White Skate *(Bathyraja spinosissima)*
☐ Roughtail Skate *(Bathyraja trachura)*

HARDNOSE SKATES (RAJIDAE)
☐ Broad Skate *(Amblyraja badia)*
☐ Big Skate *(Raja binoculata)*
☐ California Skate *(Raja inornata)*
☐ Longnose Skate *(Raja rhina)*
☐ Pacific Starry Skate *(Raja stellulata)*

ROUND STINGRAYS (UROLOPHIDAE)
- [ ] Round Stingray *(Urobatis halleri)*

WHIPTAIL STINGRAYS (DASYATIDAE)
- [ ] Diamond Stingray *(Dasyatis dipterura)*
- [ ] Pelagic Stingray *(Dasyatis violacea)*

BUTTERFLY RAYS (GYMNURIDAE)
- [ ] California Butterfly Ray *(Gymnura marmorata)*

EAGLE RAYS (MYLIOBATIDAE)
- [ ] Bat Ray *(Myliobatis californicus)*

DEVIL RAYS (MOBULIDAE)
- [ ] Manta *(Manta birostris)*
- [ ] Spinetail Mobula *(Mobula japanica)*

Chimaeras (Chimaeriformes)

SHORTNOSE CHIMAERAS (CHIMAERIDAE)
- [ ] White-spotted Ratfish *(Hydrolagus colliei)*
- [ ] Black Chimaera *(Hydrolagus sp.)*

LONGNOSE CHIMAERAS (RHINOCHIMARIDAE)
- [ ] Longnose Chimaera *(Harriotta raleighana)*

GLOSSARY

Adelphophagy A reproductive mode whereby the oldest developing embryos feed on smaller developing embryos within the uterus.

Adolescent Maturing.

Alar thorns Thorns on the outer disc of male skates.

Ambushing A surprise attack.

Amphipods Laterally compressed benthic crustaceans.

Anguilliform Eel-like.

Annelid worms Marine benthic worms.

Antitropical Found in temperate regions but absent from the tropics.

Aplacental viviparity A reproductive mode whereby the embryos do not receive nourishment directly from the female.

Barbels A slender tactile structure located on the nostrils of some sharks.

Batoids Raylike sharks.

Benthic Organisms that live on the bottom of the ocean.

Buccal The mouth cavity.

Burst speed A quick burst of speed used by some sharks to overtake prey.

By-catch The portion of the catch that is discarded.

California current The major current system flowing from north to south along the California coast.

Carapace The protective shell on the back of crustaceans.

Carnivore An animal that feeds predominately on other animals.

Carrion Dead and decaying animal flesh.

Cartilaginous A soft, light-colored skeletal material composed of chondrin.

Cephalopods The group of animals containing octopus and squid.

Cetaceans The group of animals containing whales and dolphins.

Circumglobal A worldwide distribution.

Cloaca A common opening for the digestive, excretory, and reproductive systems.

Congeners An individual related to another genera.

Conspecifics Members of the same species or population.

Continental shelf The portion of the sea floor found closest to shore out to a depth of about 200 m.

Continental slope The steep portion of the sea floor found below 200 m.

Copepod A group of crustaceans.

Cosmopolitan Wide ranging.

Crustaceans The group of animals that contains amphipods, crabs, isopods, and shrimps.

Cusp The point of the tip of a tooth.

Cusplet A small cusp.

Demersal An organism that lives on or near the bottom.

Denticles The scales found in cartilaginous fishes.

Dimorphism Occurring in two distinct forms, usually referring to morphological differences between male and female.

Disc width The distance from wing tip to wing tip in rays.

Dorsal The upper surface or back.

Echinoderms The group of animals containing sea stars and sea urchins.

Ecomorphology A term used by ichthyologists studying the morphology and natural history of cartilaginous fishes in combination with their apparent life history.

Ecosystem The complex of an ecological community functioning as a unit in nature.

Egg case An enclosed, flexible, hornlike protein that surrounds the eggs of cartilaginous fishes. In egg-laying species it is thick-walled and protects the egg.

Elasmobranch A member of the subclass Elasmobranchii.

Endemic A species that is unique to a certain area.

Epipelagic The upper 200 m of the oceanic environment.

Erect Upright.

Fauna The animal community of a region.

Filter feeding A mode of feeding whereby tiny planktonic organisms are filtered by specialized gill rakers.

Fin insertion The posterior portion of the fin base.

Fin origin The anterior portion of the fin base.

Fusiform Tapering toward each end; spindle shaped.

Gastropods A class of molluscs (Gastropoda) commonly referred to as snails.

Gestation Length of pregnancy.

Gill rakers The dense elongated denticles on the internal gill opening which filter out planktonic organisms.

Herbivore An animal that feeds predominately on plant material.

Heterocercal A caudal fin with an upper lobe longer than the lower lobe.

Holotype The original specimen from which a species is described.

Ichthyologist A scientist who studies fishes.

Imbricated Overlapping of edges.

Interdorsal The space between the insertion of the first and origin of the second dorsal fin.

Internarial space The space between the nostrils.

Intertidal The area of the coast that is exposed during low tide but submerged at high tide.

Intromitten To insert or put in.

Invertebrates The group of animals that lacks a backbone.

Isopods A group of ventrally compressed crustaceans.

Junior synonym An invalid scientific name.

Juvenile A young, immature animal.

Keels A ridge usually on the caudal peduncle or tail fin.

Lamnoids The mackerel sharks.

Long-liners A fishing method whereby a long-line with numerous baited hooks is set to catch fish.

Malar thorns Thorns that are located along the anterior disc edge in some male skates.

Mesopelagic That zone in the oceanic environment at a depth between 200 and 1,000 m.

Midwater The mesopelagic zone.

Molluscs The group of animals that includes marine snails, clams, mussels, and scallops.

Nictitating eyelid A membrane found in some *Carcharhinus* sharks that protects the eye.

Nuchal thorns Those thorns located on the back of a skate.

Nursery ground An area where pregnant females give birth and the young remain for a period of time. These areas usually offer protection from predators and are high in nutrients.

Oblique Slanting.

Oceanic The open ocean.

Ocellus An eyespot.

Omnivorous Feeds on both animals and plants.

Oophagy A reproductive mode whereby developing embryos feed on eggs within the uterus.

Oviparous A reproductive mode whereby the female deposits egg cases on the sea floor bottom.

Ovoviviparous A reproductive mode whereby the embryos are nourished by a yolk sac without any attachment to the female.

Parasitize An organism that lives and feeds on another organism to the detriment of that organism.

Peduncle The area between the second dorsal fin and the origin of the tail fin.

Pelagic Free-swimming marine organisms that are not dependent on the bottom.

Pheromone A chemical produced by an animal which serves to stimulate a behavior in another animal.

Photophore A light-producing organ.

Pinnipeds An order of marine mammals represented by seals and sea lions.

Placental viviparity A reproductive mode whereby the embryos receive nourishment directly from the female.

Plankton The microscopic organisms that passively float or weakly swim in a body of water.

Polychaete worm A group of segmented marine worms (class Polychaeta) that usually live in mud or sand bottoms.

Prey item An animal taken as food by another.

Proboscis A modified, elongated mouth found in some marine snails, used to drill a hole into the hard outer surface of their prey to extract the soft inner tissues.

Rhomboid Diamond-shaped, but with unequal adjacent sides and angles.

Saddlebars A blotch extending across the dorsal surface from one side to another.

Scapular thorns The thorns on a skate located on their shoulders.

Spiracle A respiratory pore located behind the eyes on some groups of sharks and rays.

Supraorbital sensory canal Part of the sensory system found above the eyes. An important characteristic in distinguishing *Apristurus* catsharks.

Surfline The area of the beach where the waves break.

Tail thorns Skate thorns located on their tail.

Taxonomy The classification of animals or plants according to their relationships.

Telemetric tag A tag with a radio transmitter that sends data to a distant station.

Teleosts The major group of fishes whose skeleton consists of bone.

Thermocline A transition layer in a stratified body of water.

Thornlets Small thorns.

Thorns Enlarged denticles with an erect, sharp cusp usually found on skates and bramble sharks.

Total length The straight-line distance from snout tip to tail tip.

Upwelling A seasonal oceanographic condition which brings cool, nutrient-rich waters to the surface.

Ventral The lower portion on sharks or the underside of rays.

Vertebral centra The center of the vertebrae.

Villi Fingerlike projections of the uterus in stingrays which secretes a fluid milky nutrient from the mother to the developing embryo.

Viviparous See placental viviparity.

Yolk sac placenta A form of reproduction in some ground sharks whereby nourishment is passed directly from the mother to the embryo.

Zooplankton Microscopic animals of the plankton.

MUSEUMS AND RESEARCH INSTITUTIONS

Pacific Shark Research Center, Moss Landing Marine Laboratories, 8272 Moss Landing Road, Moss Landing, CA 95039

California Academy of Sciences, Ichthyology Department, Golden Gate Park, San Francisco, CA 94118

Long Marine Lab, University of California–Santa Cruz, 100 Shaffer Road, Santa Cruz, CA 95060

Marine Science Department, University of California–Santa Barbara, Santa Barbara, CA 93106

Monterey Bay Aquarium, 886 Cannery Row, Monterey, CA 93940

Monterey Bay Aquarium Research Institute, 7700 Sandholt Road, Moss Landing, CA 95039

Natural History Museum Los Angeles County, Fish Department, Los Angeles, CA 90007

Scripps Institute of Oceanography, University of California–San Diego, La Jolla, CA 92093

Most California State Universities and University of California campuses have ichthyologists on staff. Also, the California Department of Fish and Game has marine field offices located in Eureka, Fort Bragg, Bodega Bay, Menlo Park, Monterey, Morro Bay, Santa Barbara, Los Alamitos, and La Jolla with staff that may be able to assist in identifying rare or unusual species.

REFERENCES

Ackerman, J. T., M. C. Kondratieff, S. A. Matern, and J. J. Cech. 2000. Tidal influence on spatial dynamics of Leopard Sharks, *Triakis semifasciata,* in Tomales Bay, California. *Environ. Biol. Fishes* 58:33–43.

American Fisheries Society. 1991. *Common and scientific names of fishes from the United States and Canada,* 5th ed. American Fisheries Society Special Publication, no. 20. Bethesda, Maryland

Ayres, W. O. 1854. Descriptions of *Osmerus elongatus* and *Mustelus felis. Proc. Calif. Acad. Nat. Sci.* 1:16–17.

Ayres, W. O. 1855a. Description of *Notorynchus maculatus. Proc. Calif. Acad. Nat. Sci.* 1:72–73.

Ayres, W. O. 1855b. A new species of cramp fish: *Torpedo californica. Proc. Calif. Acad. Nat. Sci.* 1:70–71.

Ayres, W. O. 1859. New fishes: *Squatina californica. Proc. Cal. Acad. Nat. Sci.* 2:25–32.

Babel, J. S. 1967. Reproduction, life history, and ecology of the Round Stingray, *Urolophus halleri* Cooper. *Calif. Fish Game Fish Bull.* 137:104.

Backus, R. H., S. Springer, and E. L. Arnold. 1956. A contribution to the natural history of the White-tip Shark, *Pterolamiops longimanus* (Poey). *Deep-sea Res.* 3:178–188.

Baduini, C. L. 1995. Feeding ecology of the Basking Shark (*Cetorhinus maximus*) relative to distribution and abundance of prey. Master's thesis, San Jose State University.

Barry, J. P., and N. Maher. 2000. Observation of the Prickly Shark, *Echinorhinus cookei,* from the oxygen minimum zone in Santa Barbara Basin, California. *Calif. Fish Game* 86(3):213–215.

Barton, O. 1948. Color notes on Pacific Manta, including a new form. *Copeia* 2:146–147.

Bedford, D. W. 1987. Shark management: A case history—The California Pelagic Shark and Swordfish fishery. In S. Cook (ed.), *Sharks: An inquiry into biology, behavior, fisheries, and their use.* Oregon State University, Corvallis, pp. 161–171.

Bedford, D. W. 1992. Thresher shark. In W. S. Leet, C. M. Dewees, and C. W. Haugen (eds.), California's living marine resources and their utilization, Sea Grant Extension Publ., pp. 49–51.

Beebe, W., and J. Tee-Van. 1941. Eastern Pacific expeditions of the New York Zoological Society. XXV. Fishes from the tropical eastern Pacific. Part 2. Sharks. *Zoologica* 26:93–122.

Blainville, H. M. D. de. 1816. Prodrome d'une distribution systematique du regne animal. *Bull. Sci. Soc. Philom.* Paris 8:105–124.

Bleeker, P. 1852. Bijdrage tot de kennis der Plagiostomen van den Indischen. *Archipel. Verh. Batav. Genoots. Kunst. Wet.* 24:1–92.

Bray, R. N., and M. Hixon. 1978. Night shocker: Predatory behavior of the Pacific Electric Ray (*Torpedo californica*). *Science* 200(4339): 333–334.

Cailliet, G. M. 1992. Demography of the central California population of the Leopard Shark (*Triakis semifasciata*). *Aust. J. Mar. Freshwater Res.* 43:183–193.

Cailliet, G. M., and D. W. Bedford. 1983. The biology of three pelagic sharks from California waters and their emerging fisheries: A review. *CalCOFI Rept.* 24:57–69.

Cailliet, G. M., L. K. Martin, J. T. Harvey, D. Kusher, and B. A. Welden. 1983. Preliminary studies on the age and growth of the Blue Shark, *Prionace glauca*, Common Thresher, *Alopias vulpinus*, and Shortfin Mako, *Isurus oxyrinchus*, sharks from California waters. Proc. Int. Workshop Age Det. Oceanic Pelagic Fishes. *NOAA Tech. Rept. NMFS* 8:157–166.

Cailliet, G. M., L. J. Natanson, B. A. Welden, and D. A. Ebert. 1985. Preliminary studies on the age and growth of the White Shark, *Carcharodon carcharias*, using vertebral bands. *Mem. S. Calif. Acad. Sci.* 9:49–60.

Castro, J. I. 1983. *The Sharks of North American Waters.* Texas A&M Univ. Press, College Station, TX.

Chang, W. B., M. Y. Leu, and L. S. Fang. 1997. Embryos of the Whale Shark, *Rhincodon typus*: Early growth and size distribution. *Copeia* 2:444–446.

Clark, E., and D. R. Nelson. 1997. Young Whale Sharks, *Rhincodon typus*, feeding on a copepod bloom near La Paz, Mexico. *Environ. Biol. Fishes* 50:63–73.

Colman, J. G. 1997. A review of the biology and ecology of the Whale Shark. *J. Fish Biol.* 51:1219–1234.

Compagno, L. J. V. 1984. *Sharks of the World. An annotated and illustrated catalogue of shark species known to date.* FAO Fisheries Synopsis no. 125, vol. 4, part 1 (noncarcharhinoids), pp. 1–250, part 2 (Carcharhiniformes), pp. 251–655.

Compagno, L. J. V. 1988. *Sharks of the order Carcharhiniformes,* pp. 1–572. Princeton, NJ: Princeton University Press.

Compagno, L. J. V. 1990. Relationships of the Megamouth Shark, *Megachasma pelagios* (Megachasmidae, Lamniformes), with comments on its feeding habits. In *Elasmobranchs as living resources: Advances in the biology, ecology, systematics, and the status of the fisheries.* H. L. Pratt, Jr., S. H. Gruber, and T. Taniuchi (eds.). NOAA Tech. Rept. (90):363–385.

Compagno, L. J. V. 1999. Checklist of living elasmobranchs. In *Sharks, skates, and rays: The biology of elasmobranch fishes.* W. C. Hamlett (ed.), pp. 471–498. Baltimore, MD: Johns Hopkins University Press.

Compagno, L. J. V. 2001. *Sharks of the world. An annotated and illustrated catalogue of shark species known to date.* Volume 2. Bullhead, mackerel and carpet sharks (Heterodontiformes, Lamniformes and Orectolobiformes), 269 pp. FAO Species Catalogue for Fishery Purposes. No. 1, vol. 2. Rome, FAO.

Compagno, L. J. V., D. A. Ebert, and M. J. Smale. 1989. *Guide to the sharks and rays of southern Africa,* 160 pp. Cape Town: Struik Publishers.

Compagno, L. J. V., F. Krupp, and W. Schneider. 1995. Tiburones. In *Guia FAO para la identificacion de especies para los fines de la pesca.* W. Fischer, F. Krupp, W. Schneider, C. Sommer, K. E. Carpenter, and V. H. Niem (eds.), Vol. 2, pp. 645–741. Pacifico Centro-Oriental. Food and Agriculture Organization of the United Nations, Rome.

Cooper, J. G. 1864. [Description of *Pteroplatea marmorata*] *Proc. Calif. Acad. Sci.* 3:112.

Crane, N. L., and J. N. Heine. 1992. Observations of the Prickly Shark (*Echinorhinus cookei*) in Monterey Bay, California. *Calif. Fish Game* 78:166–168.

Croaker, R. S. 1942. Mackerel Shark (*Lamna nasus*) taken in California. *Calif. Fish Game* 28(2):124–125.

Cross, J. N. 1988. Aspects of the biology of two scyliorhinid sharks, *Apristurus brunneus* and *Parmaturus xaniurus,* from the upper continental slope off southern California. *Fish. Bull.* 86(4):691–702.

Cuvier, G. L. C. F. D. 1817. La regne animal distribue d'apres son organisation. Tome II. Les Reptiles, les Poissons, les Mollusques et les Annelides, 532 pp. Deterville, Paris.

Daniel, J. F. 1934. *The elasmobranch fishes,* Berkeley, CA. University of California Press.

Daugherty, A. E. 1964. The Sand Shark, *Carcharhinus ferox* (Risso), in California. *Calif. Fish Game* 50(1):4–10.

Dempster, R. P., and E. S. Herald. 1961. Notes on the Horn Shark, *Heterodontus francisci,* with observations on mating activities. *Occas. Pap. Calif. Acad. Sci.* 33:1–7.

Didier, D. A. 1995. Phylogenetic systematics of extant chimaeroid fishes (Holocephali, Chimaeroidei). *Am. Mus. Novitates* 3119: 1–86.

Dumeril, A. 1865. Histoire naturelle des poissons ou ichthyologie generale. Tome Premier. Elasmobranch, Plagiostomes et Holocephales, ou Chimeres. Premiere Partie. Paris: 720 pp. Librarie Encyclopedique de Roret.

Ebert, D. A. 1985. Color variation in the Sevengill Shark, *Notorynchus maculatus* Ayres, along the California coast. *Calif. Fish Game* 71(1):58–59.

Ebert, D. A. 1986a. Biological aspects of the Sixgill Shark, *Hexanchus griseus. Copeia* 1986(1):131–135.

Ebert, D. A. 1986b. Aspects on the biology of hexanchoid sharks along the California coast. In *Indo-Pacific Fish Biology: Proceedings of the Second International Conference on Indo-Pacific Fishes.* T. Uyeno, R. Arai, T. Taniuchi, and K. Matsuura (eds.), pp. 437–449. Tokyo: Ichthyology Society of Japan.

Ebert, D. A. 1986c. Observations on the elasmobranch assemblage of San Francisco Bay. *Calif. Fish Game* 72(4):244–249.

Ebert, D. A. 1989. Life history of the Sevengill Shark, *Notorynchus cepedinaus* Peron 1807, in two northern California bays. *Calif. Fish Game* 75(2):102–112.

Ebert, D. A. 1990. The taxonomy, biogeography, and biology of cow and frilled sharks (Chondrichthyes: Hexanchiformes). Ph.D. dissertation. Grahamstown, South Africa: Rhodes University.

Ebert, D. A. 1991. Observations on the predatory behaviour of the Sevengill Shark, *Notorynchus cepedianus. S. Afr. J. Mar. Sci.* 11: 455–465.

Ebert, D. A. 1992. Other mackerel sharks. In *California's living marine resources and their utilization*. W. S. Leet, C. M. Dewees, and C. W. Haugen (eds.), pp. 55–56. Sea Grant Extension Publ.: Davis, CA.

Ebert, D. A. 1994. Diet of the Sixgill Shark *Hexanchus griseus* off southern Africa. *Afr. J. Mar. Sci.* 14:213–218.

Ebert, D. A. 2001. First eastern Pacific records of the Longfin Mako Shark, *Isurus paucus*. *Calif. Fish Game* 87(3):117–121.

Ebert, D. A., L. J. V. Compagno, and L. J. Natanson. 1987. Biological notes on the Pacific Sleeper Shark, *Somniosus pacificus* (Chondrichthyes: Squalidae). *Calif. Fish Game* 73(2):117–123.

Eckert, S. A., and B. S. Stewart. 2001. Telemetry and satellite tracking of Whale Sharks, *Rhincodon typus,* in the Sea of Cortez, Mexico, and the north Pacific Ocean. *Environ. Biol. Fishes* 60: 299–308.

Edwards, H. M. 1920. The growth of the Swell Shark within the egg case. *Calif. Fish Game* 6(4):153–157.

Eschmeyer, W. N., E. S. Herald, and H. Hammond. 1983. *A guide to the Pacific coast fishes of North America*. Field Guide 28. Boston: Houghton Mifflin.

Faber, F. 1829. *Naturgeschichte der fische Islands*. Frankfurt a M.

Feder, H. M., C. H. Turner, and C. Limbaugh. 1974. Observations on kelp fishes associated with kelp beds in southern California, 144 pp. *Calif. Fish Game Fish Bull.* 160.

Fitch, J. E., and W. L. Craig. 1964. First records for the Bigeye Thresher Shark (*Alopias supercilliosus*) and Slender Tuna (*Allothunnus fallai*) from California, with notes on eastern Pacific scombrid otoliths. *Calif. Fish Game* 50(3):195–206.

Fowler, H. W. 1941. The fishes of the groups Elasmobranchii, Holocephali, Isospondyli, and Ostariophysi obtained by United States Bureau of Fisheries Steamer Albatross in 1907 to 1910, chiefly in the Philippine Islands and adjacent seas. 879 pp. *Bull. U.S. Nat. Mus.* (100)13.

Fouts, W. R., and D. R. Nelson. 1999. Prey capture by the Pacific Angel Shark, *Squatina californica*: Visually mediated strikes and ambush-site characteristics. *Copeia* 2:304–312.

Fry, D. H., and P. M. Roedel. 1945. The shark, *Carcharhinus azureus,* in southern California waters. *Calif. Fish Game* 31(4):209.

Fusaro, C., and S. Anderson. 1980. First California record: The Scalloped Hammerhead Shark, *Sphyrna lewini,* in coastal Santa Barbara waters. *Calif. Fish Game* 66(2):121–123.

Galvan-Magana, F., H. J. Nienhuis, and A. P. Klimley. 1989. Seasonal abundance and feeding habits of sharks of the lower Gulf of California, Mexico. *Calif. Fish Game* 75(2):74–84.

Garman, S. 1880. New species of selachians in the Museum collection. *Bull. M.C.Z.* 6:167–172.

Garman, S. 1881. Synopsis and descriptions of the American Rhinobatidae. *Proc. U.S. Nat. Mus.* 3:516–523.

Garman, S. 1885. Notes and descriptions taken from selachians in the U.S. National Museum. *Proc. U.S. Nat. Mus.* 8:39–44.

Garman, S. 1906. New Plagiostoma. *Bull. M.C.Z.* 46:201–208.

Garman, S. 1913. The Plagiostoma, 515 pp. *Mem. Mus. Comp. Zool. Harvard* 36.

Garrick, J. A. F. 1982. Sharks of the genus *Carcharhinus*. Nat. Ocean. Atmosph. Admin. U.S.A., Tech. Rept., Nat. Mar. Fish. Serv. Circ. (445), 194 pp.

Gilbert, C. H. 1892. Descriptions of thirty-four new species of fishes collected in 1888 and 1889, principally among the Santa Barbara Islands and in the Gulf of California. *Proc. U.S. Nat. Mus.* 14:539–566.

Gilbert, C. H. 1915. Fishes collected by the United States fisheries steamer "Albatross" in southern California in 1904. *Proc. U.S. Nat. Mus.* 48:305–380.

Gilbert, C. H., and E. C. Starks. 1904. The fishes of Panama Bay, 304 pp. *Mem. Calif. Acad. Sci.* 4.

Gilbert, C. R. 1967. A revision of the hammerhead sharks (family Sphyrnidae). *Proc. U.S. Nat. Mus.* 119(3539):1–88.

Gill, T. 1861. Analytical synopsis of the order Squali. *Ann. Lyceum Nat. Hist.* N.Y. 7:367–408.

Gill, T. 1862. On the classification of the families and genera of the Squali of California. *Proc. Acad. Nat. Sci. Phil.* 13: 483–501.

Gill, T. 1863. On the classification of the families and genera of the Squali of California. *Proc. Acad. Nat. Sci. Philadelphia* 14: 483–501.

Gill, T. 1864. Second contribution to the selachology of California. *Proc. Acad. Nat. Sci. Philadelphia* 16:147–151.

Gill, T. 1865. Note on the family of myliobatoids, and on a new species of *Aetobatis*. *Ann. N.Y. Lyceum* 8:135–138.

Girard, C. F. 1854. Characteristics of some cartilaginous fishes of the Pacific coast of North America. *Proc. Acad. Nat. Sci. Philadelphia* 7:196–197.

Girard, C. F. 1856. Contributions to the ichthyology of the western coast of the United States. *Proc. Acad. Nat. Sci. Philadelphia* 8:131–137.

Girard, C. F. 1858. Fishes. Report of the Pacific Railroad, Washington. 10(pt. 4):372.

Gray, A. E., T. J. Mulligan, and R. W. Hannah. 1997. Food habits, occurrence, and population structure of the Bat Ray, *Myliobatis californica,* in Humboldt Bay, California. *Environ. Biol. Fishes* 49:227–238.

Gray, J. E. 1851. List of the specimens of fish in the collection of the British Museum. Part I. Chondropterygii, 160 pp. London: British Museum (Natural History).

Grover, C. A. 1972a. Population differences in the Swell Shark, *Cephaloscyllium ventriosum. Calif. Fish Game* 58(3): 191–197.

Grover, C. A. 1972b. Predation on egg cases of the Swell Shark, *Cephaloscyllium ventriosum. Copeia* 4:871–872.

Grover, C. A. 1974. Juvenile denticles of the Swell Shark, *Cephaloscyllium ventriosum:* Function in hatching. *Can. J. Zool.* 52: 359–363

Gunther, A. 1870. Catalogue of the fishes in the British Museum, Vol. 8. 549 pp. London: British Museum (Natural History).

Hanan, D. A., D. B. Holts, and A. T. Coan. 1993. The California drift gill net fishery for sharks and Swordfish, 1981–82 through 1990–91. *Calif. Fish Game Fish Bull.* 175:1–195.

Hart, J. L. 1973. Pacific fishes of Canada, 740 pp. *Fish. Res. Bd. Can. Bull.* 180.

Harvey, J. T. 1989. Food habits, seasonal abundance, size, and sex of the Blue Shark, *Prionace glauca,* in Monterey Bay, California. *Calif. Fish Game* 75(1):33–44.

Homma, K., T. Maruyama, T. Itoh, H. Ishihara, and S. Uchida. 1999. Biology of the Manta Ray, *Manta birostris* Walbaum, in the Indo-Pacific. In *Proceedings of the 5th Indo-Pacific Fish Conference, Noumea, 1997.* B. Seret and J. Y. Sire (eds.), pp. 209–216. Paris: Soc. Fr. Ichthyol.

Hubbs, C. L., and W. I. Follet. 1947. *Lamna ditropis,* new species, the Salmon Shark of the North Pacific. *Copeia,* 3:194.

Hubbs, C. L., and J. L. McHugh. 1950. Pacific Sharpnose Shark (*Scoliodon longurio*) in California and Baja California. *Calif. Fish Game* 36(1):7–11.

Hubbs, C. L., T. Iwai, and K. Matsubara. 1967. External and internal characters, horizontal and vertical distribution, luminescence, and food of the Dwarf Pelagic Shark, *Euprotomicrus bispinatus*. 64 pp. *Bull. Scripps Inst. Oceanogr.* 10.

Hubbs, C. L., W. I. Follett, and L. J. Dempster. 1979. List of the fishes of California. *Occ. Pap. Calif. Acad. Sci.* 33:1–51.

Ishiyama, R. 1958. Studies on the rajid fishes (Rajidae) found in the waters around Japan. *J. Shimonoseki Coll. Fish.* 7:193–394.

Ishiyama, R., and C. L. Hubbs. 1968. *Bathyraja,* a genus of Pacific skates (Rajidae) regarded as phyletically distinct from the Atlantic genus *Breviraja. Copeia* 2:407–410.

Johnson, A. G., and H. F. Horton. 1972. Length–weight relationship, food habits, parasites, and sex and age determination of the Ratfish, *Hydrolagus colliei* (Lay and Bennett). *Fish. Bull.* 70(2):421–429.

Jones, B. C., and G. H. Geen. 1977a. Age determination of an elasmobranch (*Squalus acanthias*) by X-ray spectrometry. *J. Fish. Res. Bd. Can.* 34:44–48.

Jones, B. C., and G. H. Geen. 1977b. Reproduction and embryonic development of Spiny Dogfish, *Squalus acanthias,* in the Strait of Georgia, British Columbia. *J. Fish. Res. Bd. Can.* 34:1286–1292.

Jones, B. C., and G. H. Geen. 1977c. Food and feeding habits of Spiny Dogfish (*Squalus acanthias*) in British Columbia waters. *J. Fish. Res. Bd. Can.* 34:2067–2078.

Jordan, D. S., and B. W. Evermann. 1896. The fishes of North and Middle America. *Bull. U.S. Nat. Mus.* No. 47(p. 1):1240 pp.

Jordan, D. S., and C. H. Gilbert. 1880a. Description of a new species of notidanoid shark (*Hexanchus corinus*), from the Pacific coast of the United States. *Proc. U.S. Nat. Mus.* 3:352–355.

Jordan, D. S., and C. H. Gilbert. 1880b. Notes on a collection of fishes from San Diego, California. *Proc. U.S. Nat. Mus.* 3:23–34.

Jordan, D. S., and C. H. Gilbert. 1880c. Description of a new ray (*Platyrhina triseriata*) from the coast of California. *Proc. U.S. Nat. Mus.* 3:36–38.

Jordan, D. S., and C. H. Gilbert. 1880d. Description of a new species of ray (*Raia stellulata*) from Monterey, California. *Proc. U.S. Nat. Mus.* 3:133–135.

Jordan, D. S., and C. H. Gilbert. 1881. Note on *Raia inornata. Proc. U.S. Nat. Mus.* 4:73–74.

Jordan, D. S., and C. H. Gilbert. 1883a. Synopsis of the fishes of North America. *Bull. U.S. Nat. Mus.* 16:1018 pp.

Jordan, D. S., and C. H. Gilbert. 1883b. Description of a new shark (*Carcharias lamiella*) from San Diego, California. *Proc. U.S. Nat. Mus.* 5(1882):110–111.

Jordan, D. S., and C. H. Gilbert. 1883c. Description of four new species of sharks, from Mazatlan, Mexico. *Proc. U.S. Nat. Mus.* 5(1882):102–110.

Joung, S. J., C. T. Chen, E. Clark, S. Uchida, and W. Y. P. Huang. 1996. The Whale Shark, *Rhincodon typus,* is a livebearer: 300 Embryos found in "megamamma supreme." *Environ. Biol. Fishes* 46:219–223.

Kato, S. 1992. Spotted chimaera. In *California's living marine resources and their utilization.* W. S. Leet, C. M. Dewees, and C. W. Haugen (eds.), pp. 197–198. Davis, CA: Sea Grant Extension Publ.

Kato, S., S. Springer, and M. H. Wagner. 1967. Field guide to eastern Pacific and Hawaiian sharks. *U.S. Fish Wildl. Serv. Circ.* 271:47 pp.

Ketchen, K. S. 1986. The Spiny Dogfish (*Squalus acanthias*) in the northeast Pacific and a history of its utilization, 68 pp. *Can.Spec. Pub. Fish. Aquatic Sci.* 88.

Klimley, A. P. 1985. The areal distribution and autoecology of the White Shark, *Carcharodon carcharias,* off the west coast of North America. *South. Calif. Acad. Sci. Mem.* 9:15–40.

Kusher, D. I., S. E. Smith, and G. M. Cailliet. 1992. Validated age and growth of the Leopard Shark, *Triakis semifasciata,* with comments on reproduction. *Environ. Biol. Fishes* 35:187–203.

Last, P. R., and J. D. Stevens. 1994. *Sharks and rays of Australia.* CSIRO Division of Fishes, Melbourne, Australia.

Lavenberg, R. J. 1991. Megamania—The continuing saga of megamouth sharks. *Terra* 30(1):30–39.

Lavenberg, R. J., and J. A. Seigel. 1985. The Pacific's megamystery—Megamouth. *Terra* 23(4):29–31.

Lay, G. T., and E. J. Bennett. 1839. Fishes. In *The zoology of Captain Beechy's voyage.* J. Richardson, N.A. Vigors, G. T. Lay, E. T. Bennett, R. Owens, J. Gray, W. Buckland, and G. B. Sowerby (eds.), pp. 41–75. London: Henry G. Bohn.

LeBouf, B. J., J. E. McCosker, and J. Hewitt. 1987. Crater wounds on Northern Elephant Seals: The Cookiecutter Shark strikes again. *Fish Bull.* 85(2):387–392.

Lowe, C. G., R. N. Bray, and D. R. Nelson. 1994. Feeding and associated electrical behavior of the Pacific Electric Ray, *Torpedo californica,* in the field. *Mar. Biol.* 120(1):161–169.

MacGinitie, G. E. 1947. Notes on the Devilfish, *Mobula lucasana*, and its parasites. *Copeia* 4:276–278.

Martin, L. K., and G. M. Cailliet. 1988a. Aspects of the reproduction of the Bat Ray, *Myliobatis californica* Gill, in central California. *Copeia* 3:752–762.

Martin, L. K., and G. M. Cailliet. 1988b. Age and growth determination of the Bat Ray, *Myliobatis californica* Gill, in central California. *Copeia* 3:762–773.

Martin, L. K., and G. D. Zorzi. 1993. Status and review of the California skate fishery. In *Conservation biology of elasmobranchs*. S. Branstetter (ed.), NOAA Tech. Rep. NMFS 115:39–52.

Matern, S. A., J. J. Cech, and T. E. Hopkins. 2000. Diel movements of Bat Rays, *Myliobatis californica*, in Tomales Bay, California: Evidence for behavioral thermoregulation? *Environ. Biol. Fishes* 58:173–182.

Mathews, C. P., and J. D. Gonzalez. 1975. Potencial pesquero y estudios ecologicos de Bahia Magdalena III. Las extistencias de rayas con especial interes a las ya aprovechades. *Ciencias Marinas* 2(1):67–72.

McCosker, J. E. 1985. White Shark attack behavior: Observations of and speculations about predator and prey strategies. *South. Calif. Acad. Sci. Mem.* 9:123–135.

Miller, D. J., and R. N. Lea. 1972. Guide to the coastal marine fishes of California. *Calif. Dept. Fish Game Fish Bull.* No. 157:235 pp.

Mollet, H. F. 2002. Distribution of the Pelagic Stingray. *Dasyatis violacea* (Bonaparte, 1832), off California, Central America, and worldwide. *Mar. Freshwater Res.* 53:525–530.

Mollet, H. F., G. M. Cailliet, A. P. Klimley, D. A. Ebert, A. D. Testi, and L. J. V. Compagno. 1996. A review of length validation methods and protocols to measure large White Sharks. In *Great White Sharks: The biology of Carcharodon carcharhias*. A. P. Klimley and D. G. Ainley (eds.), pp. 91–108. San Diego: Academic Press.

Mollet, H. F., G. Cliff, H. L. Pratt, and J. D. Stevens. 2000. Reproductive biology of the female Shortfin Mako, *Isurus oxyrinchus* (Rafinesque, 1810), with comments on the embryonic development of lamnoids. *Fish. Bull.* 98:299–318.

Mollet, H. F., J. M. Ezcurra, and J. B. O'Sullivan 2002. Captive biology of the Pelagic Stingray, *Dasyatis violacea* (Bonaparte, 1832). *Mar. Freshwater Res.* 53:531–541.

Muller, J., and Henle, F. G. J. 1839. Systematische Beschreibung der Plagiostomem. Veit, Berlin, pp. 39–102.

Myrberg, A. A. 1991. Distinctive markings of sharks: Ethological considerations of visual function. *J. Exp. Zool.* (Suppl.), 5: 156–166.

Natanson, L. J., and G. M. Cailliet. 1986. Reproduction and development of the Pacific Angel Shark, *Squatina californica,* off Santa Barbara, California. *Copeia* 4:987–994.

Natanson, L. J., and G. M. Cailliet. 1990. Vertebral growth zone deposition in Pacific Angel Sharks. *Copeia* 4:1133–1145.

Natanson, L. J., J. G. Casey, and N. E. Kohler. 1995. Age and growth estimates for the Dusky Shark, *Carcharhinus obscurus,* in the western North Atlantic Ocean. *Fish. Bull.* 93:116–126.

Neer, J. A., and G. M. Cailliet. 2001. Aspects of the life history of the Pacific Electric Ray, *Torpedo californica* (Ayres). *Copeia* 3:842–847.

Nelson, D. R., and R. H. Johnson. 1970. Diel activity rhythms in the nocturnal, bottom-dwelling sharks, *Heterodontus francisci* and *Cephaloscyllium ventriosum. Copeia* 4:732–739.

Nelson, D. R., J. N. McKibben, W. R. Strong, C. G. Lowe, J. A. Sisnero, D. M. Schroeder, and R. J. Lavenberg. 1997. An acoustic tracking of a Megamouth Shark, *Megachasma pelagios*: A crepuscular vertical migrator. *Environ. Biol. Fishes* 49:389–399.

Nishida, K. 1990. Phylogeny of the suborder Myliobatidoidei. *Mem. Fac. Fish. Hokkaido Univ.* 37(1.2):1–108.

Noble, E. R. 1948. On the recent frilled shark catch. *Science* 108 (2806):380.

Nordell, S. E. 1994. Observations of the mating behavior and dentition of the Round Stingray, *Urolophus halleri. Environ. Biol. Fishes* 39:219–229.

Notarbartolo-di-Sciara, G. 1987. A revisionary study of the genus *Mobula* Rafinesque, 1810 (Chondrichthyes: Mobulidae) with the description of a new species. *Zool. J. Linnean Soc. London* (1987) 91:1–91.

Notarbartolo-di-Sciara, G. 1988. Natural history of the rays of the genus *Mobula* in the Gulf of California. *Fish Bull.* 86(1):45–66.

Osburn, R. C., and J. T. Nichols. 1916. Shore fishes collected by the "Albatross" expedition in Lower California with descriptions of new species. *Bull. Am. Mus. Nat. Hist.* 35(16):139–181.

Peron, F. 1807. Squalus cepedianus (p. 337). In: Voyage de decouvertes aux Terres Australes.… 1800–1802. F. Peron and L. C. Freycinet, 1807–1816, Paris, 2 volumes and atlas.

Philips, J. B. 1948. Basking Shark fishery revived in California. *Calif. Fish Game* 34(1):11–23.

Plant, R. 1989. A northern range extension for the thornback, *Platyrhinoidis triseriata. Calif. Fish Game* 75(1):54.

Poey, F. 1860. Memorias sobre la historia natural de la Isla de Cuba. Viuda de Barcina, Havana, Cuba, Vol. 2, p. 335.

Quinn, T. P., B. S. Miller, and R. C. Wingert. 1980. Depth distribution and seasonal diel movements of Ratfish, *Hydrolagus colliei,* in Puget Sound, Washington. *Fish. Bull.* 78(3): 816–821.

Radovich, J. 1961. Relationships of some marine organisms of the northeast Pacific to water temperatures (particularly during 1957 through 1959). *Calif. Fish Game Fish Bull.* 112: 1–62.

Rafinesque, C. S. 1810. *Caratteri di alcuni nuovi generi e nuevi spece di animali e piante della Sicilia,* 105 pp. Palermo.

Regan, C. T. 1908. A synopsis of the sharks of the family Scyliorhinidae. *Ann. Mag. Nat. Hist.* Ser. 8 1:453–465.

Ripley, W. E. 1946. The biology of the Soupfin, *Galeorhinus zyopterus,* and biochemical studies of the liver. *Calif. Fish Game Fish Bull.* 64:93 pp.

Roedel, P. M. 1950. Notes on two species of sharks from Baja California. *Calif. Fish Game* 36(3):330–332.

Roedel, P. M. 1951. The Brown Catshark, *Apristurus brunneus,* in California. *Calif. Fish Game* 37(1):61–63.

Roedel, P. M., and W. E. Ripley. 1950. California sharks and rays. *Calif. Fish Game Fish Bull.* 75:88 pp.

Rosenblatt, R. H., and W. J. Baldwin. 1958. A review of the eastern Pacific sharks of the genus *Carcharhinus,* with a redescription of *C. malpeloensis* (Fowler) and California records of *C. remotus* (Dumeril). *Calif. Fish Game* 44(2):137–159.

Russo, R. A. 1975. Observations on the food habits of Leopard Sharks (*Triakis semifasciata*) and Brown Smoothhounds (*Mustelus henlei*). *Calif. Fish Game* 61(2):95–103.

Salazar-Hermosa, F., and C. Villavicencio-Garayzar. 1999. Relative abundance of the Shovelnose Guitarfish *Rhinobatos productus* (Ayres, 1856) (Pisces: Rhinobatidae) in Bahia Almejas, Baja California Sur, from 1991 to 1995. *Ciencias Marinas* 25(3): 401–422.

Saunders, M. W., and G. A. McFarlane. 1993. Age and length at maturity of the female Spiny Dogfish in the Strait of Georgia, British Columbia. *Can. Environ. Biol. Fishes* 38:49–57.

Sciarrotta, T. C., and D. R. Nelson. 1977. Diel behavior of the Blue Shark, *Prionace glauca,* near Santa Catalina Island. *Calif. Fish. Bull.* 75(3):519–528.

Scofield, N. B. 1920. Sleeper shark captured. *Calif. Fish Game* 6(2):80.

Scofield, W. L. 1941. Occurrence of the Tiger Shark in California. *Calif. Fish Game* 27(4):271–272.

Segura-Zarzosa, J. C., L. A. Abitia-Cardenas, and F. Galvan-Magana. 1997. Observations on the feeding habits of the shark *Heterodontus francisci* (Girard, 1854) (Chondrichthyes: Heterodontidae), in San Ignacio Lagoon, Baja California Sur, Mexico. *Ciencias Marinas* 23(1):111–128.

Seigel, J. A. 1985. The Scalloped Hammerhead *Sphyrna lewini* in coastal southern California waters: Three records including the first reported juvenile. *Calif. Fish Game* 71(3):189–190.

Seigel, J. A., and L. J. V. Compagno. 1986. New records of the Ragged Tooth Shark, *Odontaspis ferox,* from California waters. *Calif. Fish Game* 72(3):172–176.

Seigel, J. A., D. J. Long, J. M. Round, and J. Hernandez. 1995. The Tiger Shark, *Galeocerdo cuvier,* in coastal southern California waters. *Calif. Fish Game* 81(4):163–166.

Seki, T., T. Taniuchi, H. Nakano, and M. Shimizu. 1998. Age, growth, and reproduction of the Oceanic Whitetip Shark from the Pacific Ocean. *Fish. Sci.* 64(1):14–20.

Smith, S. E., and N. J. Abramson. 1990. Leopard Shark *Triakis semifasciata* distribution, mortality rate, yield, and stock replenishment estimates based on a tagging study in San Francisco Bay. *Fish. Bull.* 88(2):371–381.

Springer, S. 1940. Three new sharks of the genus *Sphyrna* from the Pacific coast of the tropical America. *Stanford Ichthyol. Bull.* 1(5):161–169.

Springer, V. G. 1964. A revision of the carcharhinid shark genus *Scoliodon, Loxodon,* and *Rhizoprionodon. Proc. U.S. Nat. Mus.* 115(3493):559–632.

Springer, V. G., and J. A. F. Garrick. 1964. A survey of vertebral number in sharks. *Proc. U.S. Nat. Mus.* 116:73–96.

Squire, J. L. 1967. Observations of Basking Sharks and Great White Sharks in Monterey Bay: 1948–1950. *Copeia* 1:247–250.

Squire, J. L. 1990. Distribution and apparent abundance of the Basking Shark, *Cetorhinus maximus,* off the central and southern California coast, 1962–1985. *Mar. Fish. Rev.* 52(2):8–11.

Starks, E. C. 1917. The sharks of California. *Calif. Fish Game* 3(4): 145–153.

Starks, E. C. 1918. The skates and rays of California, with an account of the ratfish. *Calif. Fish Game* 4(1):1–15.

Strong, W. R. 1989. Behavorial ecology of Horn Sharks, *Heterodontus francisci*, at Santa Catalina Island, California, with emphasis on patterns of space utilization. Master's thesis, California State University, Long Beach.

Talent, L. G. 1976. Food habits of the Leopard Shark, *Triakis semifasciata*, in Elkhorn Slough, Monterey Bay, California. *Calif. Fish Game* 62(4):286–298.

Talent, L. G. 1982. Food habits of the Gray Smoothhound, *Mustelus californicus*, the Brown Smoothhound, *Mustelus henlei*, the Shovelnose Guitarfish, *Rhinobatos productus*, and the Bat Ray, *Myliobatis californica*, in Elkhorn Slough, California. *Calif. Fish Game* 68(4):224–234.

Talent, L. G. 1985. The occurrence, seasonal distribution, and reproductive condition of elasmobranch fishes in Elkhorn Slough, California. *Calif. Fish Game* 71(4):210–219.

Taylor, L. R. 1972. *Apristurus kampae*, a new species of scyliorhinid shark from the eastern Pacific Ocean. *Copeia* 1:71–78.

Teshima, K., and S. Tomonaga. 1986. Reproduction of the Aleutian Skate, *Bathyraja aleutica*, with comments on embryonic development. In *Indo-Pacific fish biology: Proceedings of the second international conference on Indo-Pacific fishes, 1986*. T. Uyeno, R. Arai, T. Taniuchi, and K. Matsuura(eds.), pp. 303–309. Ichthyology Society of Japan, Tokyo.

Timmons, M., and R. N. Bray. 1997. Age, growth, and sexual maturity of Shovelnose Guitarfish, *Rhinobatos productus* (Ayres). *Fish. Bull.* 95: 349–359.

Townsend, C. H., and J. T. Nichols. 1925. Deep sea fishes of the "Albatross" Lower California expedition. *Bull. Am. Mus. Nat. Hist.* 52(1):1–20.

Tricas, T. C. 1979. Relationship of the Blue Shark, *Prionace glauca*, and its prey species near Santa Catalina Island, *Calif. Fish. Bull.* 77(1):175–182.

Tricas, T. C., and J. E. McCosker. 1984. Predatory behavior of the White Shark (*Carcharodon carcharias*), with notes on its biology. *Proc. Calif. Acad. Sci.* 43:221–238.

Ugoretz, J. K., and J. A. Seigel. 1999. First eastern Pacific record of the Goblin Shark, *Mitsukurina owstoni* (Lamniformes: Mitsukurinidae). *Calif. Fish Game* 85(3):118–120.

Valadez-Gonzalez, C., B. Aguilar-Palomino, and S. Hernandez-Vazquez. 2001. Feeding habits of the Round Stingray *Urobatis halleri* (Cooper, 1863) (Chondrichthyes: Urolophidae) from the continental shelf of Jalisco and Colima, Mexico. *Ciencias Marinas* 27(1):91–104

Van Blaricom, G. R. 1982. Experimental analysis of structural regulation in a marine sand community exposed to oceanic swell. *Ecol. Monogr.* 52(3):283–305.

Van Dykhuizen, G., and H. F. Mollet. 1992. Growth age estimation and feeding of captive Sevengill Sharks, *Notorynchus cepedianus*, at the Monterey Bay Aquarium. *Aust. J. Mar. Freshwater Res.* 43:297–318.

Varoujean, D. H. 1972. Systematics of the genus *Echinorhinus* Blainville, based on a study of the Prickly Shark, *Echinorhinus cookei* Pietschmann. Master's thesis, Fresno State College, Fresno, CA.

Villavicencio-Garayzar, C. J. 1993a. Biologia reproductiva de *Rhinobatos productus* (Pisces: Rhinobatidae), en Bahia Almejas, Baja California Sur, Mexico. *Rev. Biol. Trop.* 41(3):777–782.

Villavicencio-Garayzar, C. J. 1993b. Notas sobre *Gymnura marmorata* (Cooper) (Pisces: Dasyatidae) en Bahia Almejas, B.C.S., Mexico. *Rev. Inv. Cient.* 4(1):91–94.

Villavicencio-Garayzar, C. J. 1995a. Reproductive biology of the Banded Guitarfish, *Zapterix exasperata* (Pisces: Rhinobatidae), in Bahia Almejas, Baja California Sur, Mexico. *Ciencias Marinas* 21(2):141–153.

Villavicencio-Garayzar, C. J. 1995b. Distribucion temporal y condicion reproductiva de las rayas (Pisces: Batoidei), capturadas comercialmenta en Bahia Almejas, B.C.S., Mexico. *Rev. Inv. Cient. Ser. Cienc. Mar. UABCS* 6(1–2):1–12.

Villavicencio-Garayzar, C. J. 1996a. Aspectos poblacionales del angelito, *Squatina californica* Ayres, en Baja California, Mexico. *Rev. Inv. Cient. Ser. Cienc. Mar. UABCS* 7 (1–2):15–21.

Villavicencio-Garayzar, C. J. 1996b. The Ragged-Tooth Shark, *Odontaspis ferox* (Risso, 1810), in the Gulf of California. *Calif. Fish Game* 82(4):195–196.

Walbaum, J. J. 1792. P. Artedi Genera Pisc. *Emend. Ichthyol.* 3:535.

Walford, L. A. 1931. Northward occurrence of southern fish off San Pedro in 1931. *Calif. Fish Game* 17(4):401–405.

Walford, L. A. 1935. The sharks and rays of California. *Calif. Fish Game Fish Bull.* 45:66.

Walter, J. P., and D. A. Ebert. 1991. Preliminary estimates of age of the Bronze Whaler *Carcharhinus brachyurus* (Chondrichthyes: Carcharhinidae) from southern Africa, with a review of some life history parameters. *S. Afr. J. Mar. Sci.* 10:37–44.

Widder, E. A. 1998. A predatory use of counter-illumination by the Squaloid Shark, *Isistius brasiliensis. Environ. Biol. Fishes* 53: 267–273.

Winter, S. P. 2001. Preliminary study of vertebral growth rings in the Whale Shark, *Rhincodon typus,* from the east coast of South Africa. *Environ. Biol. Fishes* 59:441–451.

Yano, K., F. Sato, and T. Takahashi. 1999. Observation of mating behavior of the Manta Ray, *Manta birostris,* at the Ogasawara Islands, Japan. *Ichthyol. Res.* 46(3):289–296.

Yudin, K. G., and G. M. Cailliet. 1990. Age and growth of the Gray Smoothhound, *Mustelus californicus,* and the Brown Smooth-hound, *M. henlei,* sharks from central California. *Copeia* 1: 191–204.

Zeiner, S. J., and P. Wolf. 1993. Growth characteristics and estimates of age at maturity of two skates (*Raja binoculata* and *Raja rhina*) from Monterey Bay, California. In *Conservation biology of elasmobranchs,* S. Branstetter (ed.), NOAA Tech. Rep. NMFS 115:87–99.

Zorzi, G. D., and M. E. Anderson. 1988. Records of the deep-sea skates, *Raja (Ambyraja) badia* (Garman, 1899) and *Bathyraja abyssicola* (Gilbert, 1896) in the eastern North Pacific, with a new key to California skates. *Calif. Fish Game* 74(2):87–105.

Zorzi, G. D., and L. K. Martin. 1994. The batoid fishes of California—part 1: The skates. *Chondros* 5(2):1–6.

Zorzi, G. D., and L. K. Martin. 1995. The batoid fishes of California—part 2: The sawfishes, guitarfishes, and electric rays, with general notes on the taxa. *Chondros* 6(3):1–9.

INDEX

Page references in **boldface** refer to the main discussion of the species.